Green Belts

Most of us have heard of green belts – but how much do we really know about them? This book tries to separate the fact from the fiction when it comes to green belts by looking both backwards and forwards. They were introduced in the mid-twentieth century to try and stop cities merging together as they grew. There is little doubt they have been very effective at doing that, but at what cost? Are green belts still the answer to today's problems of an increasing population and ever higher demands on our natural resources?

Green Belts: Past; present; future? reflects upon green belts in the United Kingdom at a time when they have perhaps never been more valued by the public or under more pressure from development. The book begins with a historical study of the development of green belt ideas, policy and practice from the nineteenth century to the present. It discusses the impacts and characteristics of green belts and attempts to reconcile perceptions and reality. By observing examples of green belts and similar policies in other parts of the world, the authors ask what we want green belts to achieve and suggest alternative ways in which that could be done, before looking forward to consider how things might change in the coming years.

This book draws together information from a range of sources to present, for the first time, a comprehensive study of green belts in the UK. It reflects upon the gap between perception and reality about green belts, analyses their impacts on rural and urban areas, and questions why they retain such popular support and whether they are still the right solution for the UK and elsewhere. It will be of interest to anyone who is concerned with planning and development and how we can provide the homes, jobs and services we need while protecting our more valuable natural assets.

John Sturzaker, School of Environmental Sciences, University of Liverpool, UK.

Ian Mell, School of Environmental Sciences, University of Liverpool, UK.

Routledge Studies in Urbanism and the City

This series offers a forum for original and innovative research that engages with key debates and concepts in the field. Titles within the series range from empirical investigations to theoretical engagements, offering international perspectives and multidisciplinary dialogues across the social sciences and humanities, from urban studies, planning, geography, geohumanities, sociology, politics, the arts, cultural studies, philosophy and literature.

For a full list of titles in this series, please visit www.routledge.com/series/RSUC

Beyond the Networked City
Infrastructure reconfigurations and urban change in the North and South
Edited by Olivier Coutard and Jonathan Rutherford

**Technologies for Sustainable Urban Design
and Bioregionalist Regeneration**
Dora Francese

Markets, Places, Cities
Kirsten Seale

Shrinking Cities
Understanding urban decline in the United States
*Russell C. Weaver, Sharmistha Bagchi-Sen, Jason C. Knight
and Amy E. Frazier*

Mega-Urbanization in the Global South
Fast cities and new urban utopias of the postcolonial state
Edited by Ayona Datta and Abdul Shaban

Green Belts
Past; present; future?
John Sturzaker and Ian Mell

Green Belts

Past; present; future?

John Sturzaker and Ian Mell

Routledge
Taylor & Francis Group

LONDON AND NEW YORK

First published 2017 by Routledge

2 Park Square, Milton Park, Abingdon, Oxfordshire OX14 4RN
711 Third Avenue, New York, NY 10017

Routledge is an imprint of the Taylor & Francis Group, an informa business

First issued in paperback 2018

British Library Cataloguing in Publication Data
A catalogue record for this book is available from the British Library

Library of Congress Cataloging in Publication Data
Names: Sturzaker, John, author. | Mell, Ian, author.
Title: Green belts : past, present, future? / by John Sturzaker and Ian Mell.
Description: New York : Routledge, [2016] | Series: Routledge studies in
urbanism and the city | Includes bibliographical references and index.
Identifiers: LCCN 2016026581| ISBN 9781138858237 (hardback : alk.
paper) | ISBN 9781315718170 (ebook)
Subjects: LCSH: Urban ecology (Sociology)—Great Britain. |
Greenbelts—Great Britain. | Greenways—Great Britain. |
Natural areas—Great Britain. | Landscape protection—Great Britain. |
Regional planning—Environmental aspects—Great Britain. |
City planning—Environmental aspects—Great Britain.
Classification: LCC HT243.G7 S78 2016 | DDC 307.760941—dc23
LC record available at https://lccn.loc.gov/2016026581

ISBN: 978-1-138-85823-7 (hbk)
ISBN: 978-1-138-33939-2 (pbk)

Typeset in Times New Roman
by Book Now Ltd, London

This book is dedicated to Jenifer, Alice and Theofila

Contents

Figures

Tables

Acknowledgements

We would like to thank Andreas Schulze Bäing for his help in producing the maps in chapter 3.

It is not possible to list all of the many people who have encouraged, supported and guided us in our careers to date, but a few of those who have been particularly important are (listed alphabetically) Clive Davies, Simin Davoudi, Nick Gallent, Zan Gunn, Sue Kidd, John Pendlebury, Maggie Roe, Mark Shucksmith, Tim Townshend and Geoff Vigar. We thank them all.

Our thanks of course most importantly go to our families, without whose support we would not be where we are today.

Abbreviations

AONB	Area of Outstanding Natural Beauty
APUR	Atelier parisen d'urbanisme
BMRDA	Bengaluru Metropolitan Region Development Authority
CIAT	Countryside in and around Towns
CPRE	Campaign to Protect Rural England (previously Council for the Preservation of Rural England)
CQC	Countryside Quality Counts
DCLG	Department for Communities and Local Government
DEFRA	Department for the Environment, Food and Rural Affairs
DETR	Department for the Environment, Transport and the Regions
DHS	Department of Health for Scotland
DoE	Department for the Environment
GLRPC	Greater London Regional Planning Committee
HBF	Home Builders Federation (previously House Builders Federation)
LCA	Local Character Assessment
LCC	London County Council
LPA	Local Planning Authority
MHLG	Ministry of Housing and Local Government
MoEF	Ministry of Environment & Forests (India)
NCA	National Character Assessment/Area
NCR	National Capital Region (India)
NNR	National Nature Reserve
NPPF	National Planning Policy Framework
NSA	National Scenic Area
OAN	Objectively assessed need (for housing)
PPG	Planning Policy Guidance note
RMB	Renminbi (Chinese currency)
RSS	Regional Spatial Strategy
RTPI	Royal Town Planning Institute
SDD	Scottish Development Department
SoS	Secretary of State (usually used in the text in reference to the senior politician in England responsible for the planning system)
SSSI	Site of Special Scientific Interest

TCPA	Town and Country Planning Association
TTWA	Travel to Work Area
UGB	Urban growth boundary
UK	United Kingdom
UKIP	United Kingdom Independence Party
USA	United States of America
WTP	Willingness to Pay

1   Introduction

Context for this book

Green belts surround most of the major urban areas in the United Kingdom and play a major role in the planning and development of those towns and cities. They are important both in terms of the physical limits they place on urban expansion and as part of the "spatial imaginary" of town and country planning – the green belt has become shorthand for the broader mission of protecting the countryside of the UK, particularly in England. But how much do we really know about green belts in (at the time of writing) the second decade of the twenty-first century?

Green belts were established across the UK from 1955 onwards, though the London green belt had an earlier gestation in the 1930s. The country has changed deeply in many ways since then: the population has grown and become wealthier, better educated and more diverse; our communications infrastructure has developed so that more of the countryside is now within commuting range of our large towns and cities; the economy has changed, with far fewer of us employed in the primary and secondary industries that still dominated the country in the middle of the last century; and our relationship with the landscapes around urban areas has diversified as leisure and living patterns have altered. These, and many other changes, have had profound impacts on how we live our lives and how we go about working and relaxing.

These changes in the UK have affected how we view rural Britain – in the immediate post-war years, the importance of the countryside as a food resource to make us less dependent on foreign imports was emphasised. But since around the 1960s, in parallel with the changes in society noted above, our view of the countryside has evolved – we have come to see it as a space for nature rather than a productive space – and that nature is 'pure, idyllic and vulnerable, and rural landscapes are positioned as requiring protection from harmful human intervention' (Woods, 2005, p. 165).

The planning system has also changed: with the abolition first of County Structure Plans in 2004 and then of Regional Spatial Strategies in 2011, there is no strategic tier of planning in England. This means that, particularly where their boundaries are tightly drawn to exclude rural areas, some urban planning authorities are entirely "hemmed in" and have no formal means of seeking to accommodate their growing population beyond their existing boundaries. As a result, one option which might have been open in the past, of developing urban extensions or new towns beyond their boundaries, is no longer easily available; despite the government's enthusiasm for "garden cities", the planning mechanisms to deliver them are effectively missing.

This all illustrates the famous opening line of L. P. Hartley's *The Go-Between*: 'The past is a foreign country: they do things differently there.' But green belt policy, in the face of these changes, has remained effectively steadfast for sixty years. This is perhaps because the green belt is part of planning doctrine in the United Kingdom, possibly the only part of the planning system that "the man in the street" has heard of – and, it seems, he thinks that it should continue to be protected. Of those asked 'To what extent do you agree or disagree, in principle, that existing Green Belt land in England should be retained and not built on?', 64 per cent were in agreement (Ipsos MORI, 2015). Moreover, in the 2016 local elections, several candidates who stood on a platform opposing new housing development in the districts of Basildon and Rochford, close to London (Gardiner, 2016), were elected.

However, the green belt is a topic around which there is a great deal of heat and not a lot of light. The same Ipsos MORI poll in 2015 revealed that, before the interview undertaken by the researchers, 71 per cent of respondents 'know just a little, know nothing (but have heard of), [or] never heard of' the green belt. It is remarkable that so many support green belts when they know so little about them. Perhaps one reason is the constant coverage in the press. Picking one month and one broadsheet newspaper largely at random, May 2014 saw the following headlines in *The Guardian*: 'The green belt is a precious resource. We must protect it' (4th May); 'London housing crisis: who dares campaign for building on the greenbelt?' (13th May); 'The solution to the UK housing crisis? Build on the green belt' (18th May); 'Six reasons why we should build on the green belt' (21st May); 'Housing: Building on the green belt is not a solution to the shortage' (25th May).

We could therefore argue that the rhetoric of protection which surrounds green belts inhibits a more rational and evidence-based debate of their worth. We will consequently consider throughout the following chapters what role they play and how they fit into the wider planning system and ask whether they are still fit for purpose.

Planning and development in the UK

The four nations of the UK all have different planning systems, with different legislative bases and different focuses of national planning policy. They are, however, all united by several common principles, largely originating from legislation immediately after the Second World War – the Town and Country Planning Act (which applied to England and Wales) and the Town and Country Planning (Scotland) Act, both of 1947, and the 1944 Planning (Interim Development) Act (Northern Ireland).

Firstly, each system is discretionary and locally distinctive – in each local planning authority (LPA) area, development is guided by a different set of policies contained within a local plan (called local development plans in Wales and Scotland). Proposals for development are made through applications for planning permission to LPAs. Such applications are determined, by law, principally against the policies in the local plan. Crucially, however, an LPA can depart from the development plan *if material considerations indicate otherwise*. Hence the description of the planning system as discretionary – LPAs, and the elected councillors who make decisions on their behalf, can exercise their discretion to allow, or refuse, development as they see fit. This makes the UK planning systems intensely political in nature.

Secondly, in all four nations, most development is carried out by the private sector – in contrast to some countries, there is very little state-led land assembly or building. This means that the planning system is about guiding private-sector development, encouraging development of a type and in a location that the state (whether at national or local level) wishes to see and discouraging the converse. As a consequence, if the development that is allowed by the planning system is not sufficiently profitable to the private-sector development industry, it will not take place. As we will go on to see, this is a fundamentally different situation than that envisaged as the planning system took shape after the Second World War, when a far more interventionist state was a political consensus – for example, in the ten years following the war, the vast majority of new houses were built by local authorities.

Thirdly, if LPAs refuse a planning application, the applicant has a right of appeal against this decision to a more-or-less independent body (the *Planning Inspectorate* in England and Wales, the *Planning and Environmental Appeals Division* of the Scottish Government in Scotland and the *Planning Appeals Commission* in Northern Ireland).

There are other similarities between the planning systems of England, Wales, Scotland and Northern Ireland, but it is worth reflecting on one substantial difference between them, which is that England is unique in the UK, and indeed perhaps in the world, in that it has no statutory strategic land-use planning of any sort – there is no national land-use plan, nor is there any land-use plan which carries legal weight at any other level between the national and the local. There is national planning *policy*, with which local plans must accord, but this is not spatial – i.e., it contains policy prescriptions on various issues, including green belt, but it does not, for example, indicate spatial priorities for development. This lack of a statutory mechanism means that it is difficult to plan for, among other things, housing development in a strategic manner – for instance by considering whether the need and demand for housing arising in one LPA could be met in an adjoining LPA. The 2011 Localism Act introduced the *Duty to Cooperate*, which is supposed to help with this kind of above-local issue, but its success has been mixed (Jeraj, 2013).

These challenges are accompanied by a system of planning for new housing that is itself quite complicated. Before 2011 and the Localism Act, housing requirements were calculated at a national level and targets were set for each regional planning authority. The regional planning authority would then divide that target between each local planning authority (LPA). Each LPA would then have to set aside enough land to satisfy that target.

(Smith, 2016, p. 3)

Currently, LPAs must determine their own objectively assessed need (OAN) for housing, based, for example, on population projections, and in the same way as previously set aside enough land to meet this need through their local plans. The National Planning Policy Framework (NPPF) requires LPAs to publish a list of deliverable sites which provide five years' worth of supply against this OAN. If LPAs cannot identify a five-year supply of housing in this way, then their 'relevant policies for the supply of housing should not be considered up-to-date' (DCLG, 2012, p. 13). This in turn means that LPAs and the Planning Inspectorate should not use such policies to determine planning applications or appeals against decisions, but instead use the NPPF, which contains a 'presumption in favour of sustainable development' (ibid.). In short, without a five-year supply of land, LPAs have very little ability to influence the location of new housing.

The relationship between the green belt and the process through which new housing is developed is a critical one because, most immediately, a recent court case has determined that green belt policies should be considered part of the 'relevant policies for the supply of housing' which can be ignored if LPAs do not have a five-year supply of housing land (Young, 2016). More broadly, the green belt has, since its inception, been used to control urban sprawl, which, in the UK, tends to mean chiefly new housing development. As we mentioned in our introductory paragraphs, green belts were introduced across the UK, initially in England, from 1955. The NPPF notes that the broad extent of green belts in England has been established, and that 'New Green Belts should only be established in exceptional circumstances' (DCLG, 2012, p. 12). The focus of planning activity in relation to green belts, then, is in the retention, amendment or deletion of those that are currently in place. The NPPF, consistent with government planning advice since 1955, goes on to state that, 'once established, Green Belt boundaries should only be altered in *exceptional circumstances*, through the preparation or review of the Local Plan' (ibid., emphasis added). However, shortly after the publication of the NPPF, more than half of LPAs in England with green belt within their area were preparing to release/delete some of it to meet their needs for housing, prompting strong reactions from bodies such as the National Trust (LGiU, 2013). This in turn led to 'clarification' from the government that the need for housing *should not* outweigh the protection of the green belt and that, 'when planning for new buildings, protecting our precious green belt must be paramount' (DCLG *et al.*, 2014). LPAs are then expected to plan for their housing needs *within* the green belt – something that, as noted above, can be difficult if they are tightly bounded.

We have in this section summarised very briefly some aspects of planning systems in the UK (and specifically in England). More in-depth discussions of the complexities of these systems can be found in various planning textbooks – for example, Cullingworth *et al.* (2015). We have included this summary, however, because a broad understanding of these matters is necessary to appreciate both why the green belt is such a recurring topic of discussion among planning professionals in the UK and why it is so frequently controversial.

Stakeholders in the UK green belt

Attempting to map those who have an interest, or a "stake", in conversations and decisions involving the green belt in the UK is complicated: a multiplicity of individuals and groups may have an interest, more or less explicitly stated. There are, though, a small number of such (groups of) stakeholders who recur throughout the book and in whom we are particularly interested. Each of these broad groups has overlaps with others and subdivisions – for example, in terms of spatial scale or topical focus – but they can be broadly categorised as follows.

The first is politicians, at the local and national level. We observed above that the planning systems in the UK are political in nature – this is literally the case, in that decisions on the location of development are taken by politicians, although usually based on advice from professional town planners. Most such decisions are taken at the local level by elected councillors (sometimes called members of LPAs), though in some cases they can be made nationally by the government minister responsible for planning – in England the Secretary of State (SoS) for Communities and Local Government. The SoS can "recover" decisions on planning appeals usually taken by the Planning Inspectorate or can "call in" decisions on planning applications usually taken by LPAs. Whether at the local or national level, we must assume that any decision taken by politicians has "political" dimensions in addition to any other evidence considered – every politician from the SoS to an LPA councillor needs to consider the electoral bottom line or they would swiftly find themselves out of power. The views of the electorate, then, are of prime importance, as we return to consider below.

The second group of stakeholders is planners, individually and in groups. As we will discuss in subsequent chapters, it has been argued that planners themselves *like* the green belt as a policy because, among other things, it is simple – clarity in policy-making is a powerful attraction. Ever since the formal introduction of green belts in 1955, some planners have taken the view that they were the "'very *raison d'être*'" of the planning system (cited in Amati, 2008, p. 6), and bodies representing planners such as the Royal Town Planning Institute (RTPI) regularly comment on the green belt (cf. RTPI, 2015). As noted above, while decisions on planning are made by elected politicians, the latter are advised by planners, who are also employed by developers and landowners (see below) and so have a significant influence on how green belts are treated by politicians and the development industry alike.

The third group of stakeholders are government departments and LPAs, the balance of power between which in relation to green belt has shifted at various times. The government department currently in charge of planning is the Department for Communities and Local Government (DCLG), and it is of course not some neutral or independent body; it is comprised of civil servants, including planners, and elected politicians, including the SoS – much like LPAs. The decisions taken by the DCLG and LPAs are therefore influenced by both political and professional perspectives. It has been ever thus. As we shall see in Chapter 2, there were differences of opinion about the green belt in the 1950s between ministers and civil servants in one of the DCLG's predecessors, the Ministry of Housing and Local Government (MHLG).

The fourth group of stakeholders consists of developers and landowners, each with different specific interests as regards the short vs. the long term and the extent to which they think green belt land should be released for development, but they can be categorised broadly as being pro-development and in favour of some degree of loosening of planning restrictions. It is a mistake, however, to assume that most of this group would favour the abandonment of the planning system or very large-scale land release – research shows that developers welcome the certainty brought by the planning system (RTPI & Arup, 2015) – and simple economics suggests that, if the supply of developable land suddenly increased dramatically, the price of such land would fall. Most landowners or developers would therefore like to see their own land or development given planning permission, but not those belonging to others! There have been criticisms about the influence exercised by pro-development groups such as the Home Builders Federation (HBF) in the development of national and local planning policy (Vigar *et al.*, 2000), to which we will return in subsequent chapters.

The fifth group is made up of lobbying organisations and think-tanks, which have a strong influence on planning policy relating to green belts. This applies to both the anti-green belt perspective, where neoliberal think-tanks such as the Adam Smith Institute or Policy Exchange are identified as having a forceful impact on the policy of the current Conservative-led government (Geoghegan and Millar, 2013), and the pro-green belt position, where the Campaign to Protect Rural England (CPRE) has long been singled out as having a disproportionate influence (Pennington, 2000; Murdoch and Lowe, 2003). These two examples illustrate that green belts in the UK are most frequently discussed in oppositional terms, as a form of dichotomy between pro- and anti-development voices. We hope in this book to approach things from a more neutral perspective.

The final group we wish to mention is the most diverse of all – the "general public". As noted above, opinion polls consistently suggest strong support for the green belt. This support naturally informs the stance of politicians, and it would be a brave individual indeed who ignored public opinion. However, public support for green belt is held despite an apparent lack of understanding about what the green belt *does* or, indeed, what it *is* – and, perhaps more fundamentally, despite widely held incorrect assumptions about the scale of built development in the UK. A survey undertaken for a 2006 report on planning found that '54 per cent of respondents thought that around half or more of all land in England is developed, while only 13 per cent believed that less than a quarter is developed' (Barker, 2006, p. 43). The actual figure is around 10 per cent, so there is a significant gap between perceptions and reality. It is perhaps unsurprising that many of us are opposed to the loss of green belt when we believe that so much of the country is developed. Feeding into the keenness on the part of the English to protect the green belt from development (and, as we shall discuss in Chapter 2, they do appear to have a particularly strong view on this compared with those in the other UK nations) is the strong streak of romanticism and pastoralism that runs through English literature and continues to hold such sway over public opinion and political rhetoric. As others have argued, the English have a particular obsession with "the countryside", equating it with providing a "rural idyll" that is now, and arguably always has been, at some distance from reality (Matless, 1998; Woods, 2005). This feeds into what has been called 'the preservation paradox' (Murdoch and Lowe, 2003), whereby levels of support for the green belt and other policies remain high while most people express a preference for low-density suburban or rural living – positions that are irreconcilable.

Considering how these groups of stakeholders affect, and are affected by, the green belt and debates upon it, is a recurring theme of this book. There are other themes, specified in the next section.

Focus and structure of this book

Our geographical focus in this book is on the UK and, within that, specifically England. This is for several reasons – firstly, the English green belts are the most mature, in policy terms, and the most contentious. Some of our previous work (Sturzaker, 2010) has concentrated on the sometimes vituperative debate that can greet development proposals in the English countryside. While this debate is by no means wholly an English issue (see Clifford and Warren, 2005, for a discussion on the issue around St Andrews in Scotland), the extent of the English green belt is by far the largest among the four nations within the UK, since it surrounds all the largest metropolitan areas, which means there are more frequent discussions/debates/arguments about its deletion. So our discussion will more frequently reference the English context and English examples, but we will also draw upon the other nations of the UK to support our arguments. We have also included, in Chapter 5, comparisons with green belts (and other similar policies) in other countries, both to illustrate how widely the concept is used and as part of our search for alternatives.

Our "analytical" focus varies in some respects from chapter to chapter. We are interested in the green belt from a number of perspectives – spatially, in terms of the amount of land it covers; qualitatively, in terms of what type and quality of land is so designated; and politically, in terms of the various stakeholders who are interested in the green belt and its development for a wide range of reasons (see above). Throughout the following chapters we will return to a number of key themes to frame the current debates surrounding green belts. Each theme is, we feel, as relevant today as it was in 1955 and, if viewed with an analytical lens, helps us to understand green belts physically, socially and politically. We have identified four primary themes which will be covered.

1   *Time* – The policy position on green belts has remained almost unchanged since 1955. This provides a temporal window to assess the process of development, ongoing protection and debate in the UK. Looking at the ways in which green belts are discussed against changes in socio-demographics, personal and communal mobility, and the changing political landscape will provide scope to reflect upon the diversifying socio-political aspirations of green belt policy and practice.

2   *Spatial diversity and composition* – Green belts are a constant in the minds of many people, yet the understanding of where they are, their size and what functions they perform are more fluid. By reflecting on how the rhetoric differs from the physical realities, the following chapters will look to assess "fact" and "fiction" and develop a more rounded understanding of green belts.

3   *Politicisation of the landscape* – Over the course of the last fifty years there has been a shift in our understanding and valuation of rural or pastoral landscapes. However, although society has changed, our views on green belts appear to have remained rigid. This has led to conflicts arising when and where development (particularly housing and transport infrastructure) is proposed. This also illustrates how the changing socio-economic characteristics of the UK population are impacting on how green belts are viewed.

4   *Plus ça change, plus c'est la même chose* – The more things change, the more they stay the same. With green belts, we suggest that, the more things change in society, the more green belts stay the same!

The book consists of six chapters. Chapter 2 provides a chronological history of green belts, from the initial propositions to limit city size in relation to ancient India and Rome, through attempts by monarchs, including Queen Elizabeth I, to control the growth of London, to the origins of modern green belts in the nineteenth century. It traces the development of government and non-governmental thinking on urban expansion leading to the creation of the London green belt in the 1930s, the 1947 Town and Country Planning Act, and the 1955 circular which encouraged all English LPAs to consider introducing green belts. Moving through the second half of the twentieth century, the chapter presents a picture of almost continuous support for green belts from successive governments and brings us up to date with a review of the latest government policy. It also includes summaries of the separate evolution of green belts in Scotland, Wales and Northern Ireland.

Chapter 3 addresses the lack of systematic analysis of the impacts of green belts by reviewing a wide range of evidence on the topic. The temptation is to focus on green belts themselves and the rural areas beyond them, which is also the case here, though the chapter will importantly consider their impacts on the cities and towns encircled by them. We examine issues such as house prices, urban sprawl, built density and travel patterns. We also speculate, through the use of "counterfactuals", what the country might look like today if green belts were not in place. We look at the varied impacts of green belts on different components of "sustainability" and try to explore whether they have had a net positive or negative effect on the UK, both in utilitarian terms and considering the balance between richer and poorer groups in society.

Chapter 4 analyses the characteristics of the green belts in the UK today – their size, location, geographical spread and mix of land use – and compares these "facts" with the perceptions of the public and other commentators drawn from opinion surveys and press reports. One of the main arguments that will be expanded upon is our understanding of the characteristics of green belts. This reflects on the different perceptions people hold of green belts – spatially, conceptually and socio-economically – and goes on to challenge the established narrative of them as an essential conservation/preservation policy. It also asks whether the counter-argument that the land allocated to green belts is proportional to its value remains valid when the availability of housing sites is becoming increasingly restricted. Green belts will be examined to question whether the *value* and *quality* arguments attached to these designations are still credible. This chapter will ask whether the notion that green belts are located in areas of high-quality landscape character is (a) correct and (b) a valid reason to limit development. A number of examples of landscape quality/character assessments will be used to evaluate whether green belts are protecting high-quality landscapes or low-quality urban-fringe and farmland areas.

Chapter 5 provides a series of case studies of the implementation of green belts, and similar policies, in other countries around the world. The purpose of these case studies is to assess how green belts are used in other jurisdictions and to identify key similarities and differences between practice in the UK and that in these other countries. We look at policies in Asia (South Korea, China and India), Australia, Europe (Germany, the Netherlands, Italy and France) and North America (Toronto, Oregon, Boulder and California). Common themes emerge and some lessons for the UK can be identified.

Chapter 6 discusses alternatives to green belts for cities in the UK. It looks at different delivery/management options that might retain their important elements but also offer more flexibility, recognising that the UK today is very different than it was when green belts were introduced. The aim is to postulate whether alternatives can be identified that would retain the spatial dimension of green belts but act more responsively to the changing needs of urban and rural communities. Each of the options proposed suggests that, if we move beyond the current polarised discussion, more appropriate forms of landscape management could be utilised.

Chapter 7 concludes the book by asking "Where next?" Bearing in mind the discussion in the preceding six chapters, what is the likely future for green belts in the United Kingdom? Green belts retain strong political and public support, so what appetite is there for change? One issue of notable importance is the inescapable fact that the country, and its capital, has changed beyond all recognition since green belts were introduced. The planning system has similarly changed – one might argue that, without some form of strategic planning (abolished by the government in 2011), it is very difficult to have a properly implemented green belt policy, as tightly bound cities will find it difficult to expand to meet their housing needs. We reflect upon these issues and more. Finally, we bring the threads of the book together by returning to the "?" which concludes the title of the book: can and will green belts continue in their current form?

Our methods

In writing this book we have made use of several sources of data. Principal among them is a comprehensive review of secondary material – books, journal articles and reports that take green belts or related topics as their focus, together with press coverage of that source material. Much of this comprises opinion pieces on green belts, which we use to illustrate the different perspectives discussed, but some of it makes use of empirical data – statistics on the nature of the green belt itself or on changes in house prices, for example.

We have also gathered our own primary data on the green belt in the North West of England – we work in Liverpool, one of the cities surrounded by the north-west green belt, which we have found to be an excellent and, we believe, broadly representative example of green belts in the UK. We draw upon this primary research mainly in Chapters 3, 4 and 6.

One further point of reference for the framing of the subsequent discussions is that both authors have spent a significant amount of their academic and practitioner lives working either in or around England's green belts. John Sturzaker grew up in a village within the Lancashire green belt and has worked as a policy and development management officer for various local authorities in England containing green belt. Ian Mell worked and researched landscape issues across the length of the East of England and was previously a development officer in East Cambridgeshire, where he dealt with strategic development issues in and around the Cambridgeshire green belt. All of these experiences have helped to provide both authors with first-hand experience of the complexities of planning for, and in, the green belt.

A brief note on capitalisation

For the sake of simplicity, unless directly quoting others, we use lower-case letters throughout this book when referring to the green belt. Opinions differ as to whether this is the correct approach – government policy tends to capitalise the term, as in Green Belt; others capitalise the Green but not the belt; and yet others run the words together as in greenbelt (or Greenbelt!).

Summary

Green belts are at the same time very simple and extremely complex. Within policy–practice debates there is simultaneous support and opposition for their preservation. In spite of the significant socio-economic, environmental and policy changes since the 1950s, green belt policy at the national and, as a consequence, the local level has remained largely unchanged across the UK. This book sets out to address this apparent anomaly, in part by asking a series of questions: In this radically changed context, do green belts remain fit for purpose? Do we even agree on what that purpose is? Is the time ripe to reconsider the role of green belts and to challenge some of the founding principles of the British planning system?

While we do not suggest that we have the answers to these questions, we consider the discussions outlined here to comprise a broadly comprehensive review of green belt knowledge. Each chapter will build on the key themes of understanding temporal, spatial and policy–practice changes in the UK, and further afield, to debate whether it is time for a change in approach.

We end this introductory chapter with a quotation from a man who, though we both met him only a small number of times, inspired us both with his immense knowledge and the lightness with which he carried it – the late Professor Sir Peter Hall. Sir Peter, along with a number of his colleagues, undertook a landmark review of the English planning system in the late 1960s and early 1970s. Published as *The Containment of Urban England* (in two volumes), this work was, and is, unsurpassed in terms of its attention to detail and comprehensiveness. Sir Peter and his colleagues concluded, as their title suggests, that the planning system, with green belts as a central policy mechanism, had resulted in the physical *containment* of England's towns and cities – as indeed it was intended to do. But what it had not done, which it was also intended to do, was match this containment with the provision of new urban areas beyond the green belt. The authors found that this had resulted in inequitable consequences, with poorer urban dwellers, particularly those in public or private rented housing, bearing the costs, and wealthier people, particularly those in rural areas, reaping the benefits. We aim to assess, in our own small way, whether this remains the case and to reflect upon their following profound words:

> None of this was in the minds of the founding fathers of the planning system. They cared very much for the preservation and the conservation of rural England, to be sure. But that was only part of a total package of policies, to be enforced in the interests of all by beneficent central planning … Somewhere along the way, a great deal was lost, a system distorted and the great mass of the people betrayed.
> (Hall *et al.*, 1973, p. 433)

References

Amati, M. (2008) Green belts: a twentieth-century planning experiment, in M. Amati (ed.), *Urban Green Belts in the Twenty-first Century* (Aldershot: Ashgate), pp. 1–17.

Barker, K. (2006) *Barker Review of Land Use Planning: Final Report – Recommendations* (London, HMSO).

Clifford, B. P., and Warren, C.R. (2005) Development and the environment: perception and opinion in St Andrews, Scotland, *Scottish Geographical Journal*, 121(4): 355–84.

Cullingworth, B., Nadin, V., Hart, T., Davoudi, S., Pendlebury, J., Vigar, G., Webb, D., and Townshend, T. (2015) *Town and Country Planning in the UK* (15th ed., London: Routledge).

DCLG (2012) *National Planning Policy Framework* (London: Department for Communities and Local Government).

DCLG, Lewis, B., and Pickles, E. (2014) Councils must protect our precious green belt land, Press release, https://www.gov.uk/government/news/councils-must-protect-our-precious-green-belt-land.

Gardiner, J. (2016) Why local elections could be influenced by housing hostility, *Planning*, 22 April 2016, p. 8.

Geoghegan, J., and Millar, S. (2013) *Power 100*, http://www.planningresource.co.uk/article/1173618/power-100 [a list of the 100 most influential people in the sector].

Hall, P., Gracey, H., Drewitt, R., and Thomas, R. (1973) *The Containment of Urban England*, Vol. 2 (London: Allen & Unwin).

Ipsos MORI (2015) *Attitudes towards Green Belt land: A Study for the Campaign to Protect Rural England*, https://www.ipsos-mori.com/researchpublications/researcharchive/3611/Attitudes-towards-Green-Belt-land.aspx.

Jeraj, S. (2013) True localism or selfish politics? Why the duty to cooperate is failing, The Guardian, 28 January, www.theguardian.com/local-government-network/2013/jan/28/hertfordshire-councils-duty-to-cooperate.

LGiU (2013) Government planning policy puts England's green belts at risk, suggests new research, Press release, Local Government Information Unit, www.lgiu.org.uk/news/government-planning-policy-puts-englands-green-belts-at-risk-suggests-new-research/.

Matless, D. (1998) *Landscape and Englishness* (London: Reaktion Books).

Murdoch, J., and Lowe, P. (2003) The preservationist paradox: modernism, environmentalism and the politics of spatial division, *Transactions of the Institute of British Geographers*, 28(3): 318–32.

Pennington, M. (2000) *Planning and the Political Market: Public Choice and the Politics of Government Failure* (London: Athlone Press).

RTPI (2015) *Building in the Green Belt? A Report into Commuting Patterns in the Metropolitan Green Belt* (London: Royal Town Planning Institute).

RTPI and Arup (2015) Investing in delivery: how we can respond to the pressures on local authority planning, www.rtpi.org.uk/knowledge/research/projects/national-and-regional-research-projects/investing-in-delivery/.

Smith, L. (2016) *Planning for Housing*, House of Commons Library Briefing Paper no. 03741, (London: House of Commons).

Sturzaker, J. (2010) The exercise of power to limit the development of new housing in the English countryside, *Environment and Planning A*, 42(4): 1001–16.

Vigar, G., Healey, P., Hull, A., and Davoudi, S. (2000) *Planning, Governance and Spatial Strategy in Britain: An Institutionalist Analysis* (London: Macmillan).

Woods, M. (2005) *Contesting Rurality: Politics in the British Countryside* (Aldershot: Ashgate).

Young, C. (2016) Clarity on 'relevant policies for the supply of housing'?, *Local Government Lawyer*, 17 March, www.localgovernmentlawyer.co.uk/index.php?option=com_content&view=article&id=26324%3Aclarity-on-relevant-policies-for-the-supply-of-housing&catid=63&Itemid=31.

2   A history of green belts in the UK

Introduction

This chapter presents a chronological history of green belts in the United Kingdom (with some discussion of other locations as necessary to tell this "story"). It draws upon a wide range of published material on the topic, including books, newspapers, journals, reports and government policy pronouncements. Some of these publications have attempted more or less comprehensive reviews of the state of green belts at that time (for example, Mandelker, 1962), others have focused on one period or facet of green belt policy (for example, Foley, 1963; Amati and Yokohari, 2007), but all are valuable. Much of this material has concentrated principally on England, for similar reasons to those we identified in Chapter 1 – compared to the other nations of the UK, the green belts of England have the longest history in terms of both policy and implementation, and they have been the most controversial. Similarly, London is the focus of a substantial proportion of our sources, which itself is due to the pre-eminence of London as the largest city in the UK and, consequently, the largest example of the perceived problems of uncontrolled urban growth. Given the relative scarcity of publications relating to Wales, Scotland and Northern Ireland, the history of green belts in those nations is covered in a dedicated section towards the end of the chapter.

There are two main thematic groups of sources – firstly, the changing attitude of government towards green belts, expressed formally through policy or informally through comments from ministers, and, secondly, the opinions of what we will call the "commentariat" – academics, journalists, planners, housebuilders and campaigners. We do not in this chapter discuss in detail how, for example, the characteristics of green belt land have changed over time or the impacts of green belts. However, these and other issues have influenced what other writers have discussed, so mention is occasionally made of them. Another topic that recurs is the aim/objective/purpose of green belts. In Chapter 3 we assess what effects green belts have had in the UK. Here, therefore, we do not explicitly set out to assess from a normative perspective whether green belts have been *effective*, partly because, as we will go on to discuss, their aim/objective/purpose has, from their early days, been unclear, contested and variable. Where these are referred to in our sources, we have sought to highlight them in what follows, and we summarise the changing ideas about the purpose of green belts at the end of this chapter.

Historical examples of and writing about green belts

The credit for devising the concept of the green belt is often given to planners and others in the UK in the late nineteenth century. While the detailed development of green belts can be ascribed to this origin, understanding any historical antecedents is important, including those in other parts of the world. This demands a degree of imagination and linguistic interpretation, as, while scholars have found evidence of policy and practice related to the limiting of urban growth and the use of a space of undeveloped land of some sort around towns and cities for many years, the very term "green belt" was not widely used before the twentieth century. It is of course important to try and distinguish between the sort of *de facto* green belt that would emerge as a consequence of cities being walled for defensive purposes and a deliberate limiting of urban size coupled with the conscious establishment of some kind of stretch of land that would be protected from development.

In that context, scholars have found evidence of approaches similar to green belts in ancient India (Clapp, 1971) and Greece (Osborn, 1946). The Romans used the term *ager effatus*[1] to describe an area of land outside and beyond the city walls, and Osborn is one of several writers to refer to the Bible and mention of what appears to be a green belt in the Book of Numbers (verse 35):

> *Cities for the Levites*
>
> In the plains of Moab by the Jordan at Jericho, the Lord spoke to Moses, saying: Command the Israelites to give, from the inheritance that they possess, towns for the Levites to live in, you shall also give to the Levites pasture lands surrounding the towns. The towns shall be theirs to live in, and their pasture lands shall be for their cattle, for their livestock, and for all their animals. The pasture lands of the towns, which you shall give to the Levites, shall extend a thousand cubits all around.
>
> (New Revised Standard Version, 1989)

A cubit was a length of measurement from the elbow to the fingertip. Most sources suggest 1 cubit to be equivalent to between 40 and 50 centimetres, so on that basis the pasture lands referred to in this extract extend for up to 500 metres from the town walls.

A slightly more modern reference from literature is that in Sir Thomas More's *Utopia*, which Osborn thought 'closely approaches [Ebenezer] Howard's garden city pattern' (1946, p. 172), including as it did strict limits on the size of cities and belts of open land around them. More advocated some form of birth control, specifying that extended families should have between ten and sixteen children of the age of fourteen years or thereabout ... But if chance be that in the whole city the store increase above the just number, therewith they fill up the lack of other cities. But if so be that the multitude throughout the whole island pass and exceed the due number, then they choose out of every city certain citizens, and build up a town under their own laws in the next land where the inhabitants have much waste and unoccupied ground.

(More, [1556] 1999, p. 142)

At around the same time that More was writing, Queen Elizabeth I issued a proclamation in 1580 banning any further house-building in London. She "'doth charge and strictly command all manner of persons ... to desist and forbear from any new buildings of any house or tenement within three miles from any of the gates of the said city of London ...'" (cited in Anonymous, 1956, pp. 68–9), in order to ensure a plentiful supply of food and limit the spread of the plague (MHLG, 1962). Further attempts to limit the growth of London were made by King James I at the beginning of the seventeenth century, Oliver Cromwell in 1656, and Christopher Wren after the Great Fire of London in 1666 (Loftus Hare, 1937; MHLG, 1962). In 1661 the writer John Evelyn wrote to King Charles II suggesting that

all the land that surrounds the city, particularly in the East and South-West, is made into plots of ground of about twenty, thirty, or forty acres. These plots would be separated from each other by fences and would be made into plantations, around a hundred and fifty foot deep ... The fenced-off areas would be diligently looked after and elegantly kept. They would be planted with shrubs that would produce sweet-smelling and beautiful flowers and fill the air with their fragrant smell.

(Evelyn, [1661] 2011, p. 50)

This proposal was not taken up, and other mechanisms were evidently ineffective, as the rate of population growth of London, as well as other towns and cities in the country, continued to grow, with the period 1800 to 1900 being particularly striking.

The nineteenth century

The population of London expanded from approximately 1 million to over 6 million during the nineteenth century, as the United Kingdom became an urbanised society. During the same period the proportion of the UK population living in settlements of more than 5,000 people increased from 19 per cent to 67 per cent (Bairoch and Goertz, 1986). The industrial revolution and an agricultural depression in the last quarter of the century (Hall *et al.*, 2003) were behind this shift.

The rapidity of this urbanisation, the importance given to economic growth and the power held by landowners meant that in London and other cities much existing housing was demolished to make way for industrial development, and the increased urban population was forced into very high-density and inferior slum housing (Hall *et al.*, 2003). Unsurprisingly, poor health and disease were rampant in these slums, leading to a wave of disquiet from commentators, the public and politicians. Opinions differ as to the extent to which this concern was for philanthropic or selfish reasons, as the government was concerned, particularly by the early twentieth century, about the possibility of a communist revolution and the poor quality of recruits to the army (Short, 1982; Holmans, 1987). Regardless of the origin of this concern, it was very real: the 1842 *Report on the Sanitary Conditions of the Labouring Population of Great Britain*, by Sir Edwin Chadwick, was followed by the *Royal Commission on the State of Large Towns* of 1844–5 and the *Royal Commission on the Housing of the Working Classes* of 1884. All of these described the appalling condition of housing in many working-class areas.

At the same time others bemoaned the state of London and other cities. William Cobbett habitually referred to London as 'the great wen', and stated that 'The dispersion of the wen is the only real difficulty that I see in settling the affairs of the nation and restoring it to a happy state' (Cobbett, 1885, p. 52). Lord Rosebery, Chairman of London County Council in 1891, is cited by Ebenezer Howard as uttering the following words, in our view extraordinarily given his position:

> There is no thought of pride associated in my mind with the idea of London. I am always haunted by the awfulness of London. Sixty years ago a great Englishman, Cobbett, called it a wen. If it was a wen then, what is it now? A tumour, an elephantiasis sucking into its gorged system half the life and the blood and the bone of the rural districts.
>
> (Howard, 1898, p. 3)

But what was the solution to these problems? The writer and Member of Parliament James Silk Buckingham is credited by many, including Howard, with proposing the first "modern" green belt. In 1849, perhaps inspired by the ideas in Sir Thomas More's *Utopia*, he devised a model town surrounded by a green belt with further towns beyond it as necessary (Osborn, 1946; Hall *et al.*, 2003). These ideas were promoted by, among others, John Ruskin, who in a lecture argued that providing lodging for people required

> thorough sanitary and remedial action in the houses that we have; and then the building of more, strongly, beautifully, and in groups of limited extent, kept in proportion to their streams, and walled round, so that there may be no festering and wretched suburb anywhere, but clean and busy street within and the open country without, with a belt of beautiful garden and orchard round the walls, so that from any part of the city perfectly fresh air and grass and sight of far horizon might be reachable in a few minutes' walk.
>
> (Ruskin, [1868] 2008, p. 105)

While cities in the UK, with their problems of overcrowding and ill health, were in need of remedial action, the green belt concept was also exported to the British Empire, where new cities were being built. "Park belts" were deliberately planned for various cities in Australia and New Zealand in the mid-nineteenth century. The New Zealand Company, the private business which colonised much of that country, wrote in its instructions to Captain Smith, the man who founded the city of Wellington:

> It is desirable that the whole outside of the town, inland, should be separated from the county sections by a broad belt of land, which you will declare that the Company intends to be public property, on condition that no buildings be ever erected upon it.
>
> (Osborn, 1946, p. 177)

A similar concept was part of the plan for Adelaide (Hall *et al.*, 2003).

All these examples were evidently part of the inspiration for the work of Ebenezer Howard, identified by most observers as the proponent of the first fully developed and costed proposal for 'country belts permanently reserved against the expansion of the towns which they adjoin and surround' (Osborn, 1946, p. 167).

To-Morrow: A Peaceful Path to Real Reform

Howard's book, published in 1898 under the above title and reissued in 1902 as *Garden Cities of To-Morrow*, is, as the latter title suggests, best known for its imagining of garden cities – new settlements that would combine the best, and omit the worst, of what Howard described as the 'Magnets' of *Town* and *Country* – in the new configuration of a town–country magnet. The town–country magnet would address the lack of employment in the countryside, a result of the agricultural depression mentioned above, along with the lack of infrastructure in rural areas; at the same time it would avoid the problems inherent in the cities of the time – overcrowding, poor-quality air and water, and high land prices.

Although some have criticised Howard for his naivety and his tendency to gloss over problems such as poverty and drunkenness (Simmie, 1974), his ideas remain fundamental to the education of planners around the world, and his book has been described as 'almost without question the most important single work in the history of modern town planning' (Hall *et al.*, 2003, p. 1). There are many reviews of the specific contributions made by Howard's ideas to town planning (see, for example, Hall and Ward, 1998), and we do not propose to go into their detail here, but it is worth identifying the key components of his garden cities as they are relevant to green belts.

Perhaps the most important is what Howard meant by a *garden city*. Frederic Osborn, his 'disciple and faithful lieutenant' (Hall *et al.*, 2003, p. 1), believed that the garden city, in Howard's terms, 'means as much "a city *in* a garden" as "a city *of* gardens"' (Osborn, 1946, p. 167), with the gardens being, in effect, a green belt. This is illustrated by one of Howard's famous diagrams – No. 2, 'The Garden City' (Howard, 1898) features a great deal of green space within the boundary of the city, the 'Circle Railway'; the most striking feature of the diagram is that the city is tightly bounded, at 1,000 acres (405 hectares), approximately 1.5 miles across, with a population of around 30,000, and surrounded by 5,000 acres (2,235 hectares) of agricultural land. All of this land, urban and rural, would be in the ownership of 'four gentlemen of responsible position and of undoubted probity and honour, who hold it … in trust for the people of the Garden City' (ibid., p. 13). This common ownership was the principal means by which the use of land would be controlled.

But 'Howard advocated containment only as an element in planning a city *in a state of constant growth*' (Hall *et al.*, 1973b, p. 45, emphasis added). The garden city was not intended to be an isolated town; rather, it would form one of a

> cluster of towns … so designed that each dweller in a town of comparatively small population is afforded … the enjoyment of easy, rapid, and cheap communication with a large aggregate of the population … and … may dwell in a region of pure air and be within a few minutes' walk of the country.
>
> (Howard, 1898, p. 131)

This polycentric group of "social cities" was illustrated in the diagram shown in Figure 2.1, which was omitted from later editions. Hall *et al.* (2003, p. 7) believe this omission is why 'most readers have failed to grasp the vital fact that Social City, not the isolated Garden City, was Howard's vision.' What later editions did contain, however, was an explanation of what would happen if the original garden city reached its "target" population of 32,000:

> How shall it grow? How shall it provide for the needs of others who will be attracted by its numerous advantages? Shall it build on the zone of agricultural land which is around it, and thus for ever destroy its right to be called a 'Garden City'? Surely not … It will grow by establishing – under Parliamentary powers probably – another city some little distance beyond its own zone of 'country', so that the new town may have a zone of country of its own.
>
> (Howard, [1902] 1951, pp. 140–42)

This new city, though administratively separate, should be close enough, Howard thought, to be part of the same 'community', none of them being more than 10 miles from each other. The end result would mean that 'the resulting landscape would be quite unlike what many people today think is the epitome of good planning: it would be a cluster of small new towns with very narrow green belts, perhaps only a mile or two wide' (Hall, 1974, p. 4). It is worth the reader remembering this conception of green belts throughout the rest of this book, as it was very quickly superseded. Howard's broader ideas, however, had an immediate impact and continue to exert an influence today.

The early years of the twentieth century

The most concrete example of this impact was the construction, 40 miles north of London, of Letchworth, the first garden city, which began in 1903,

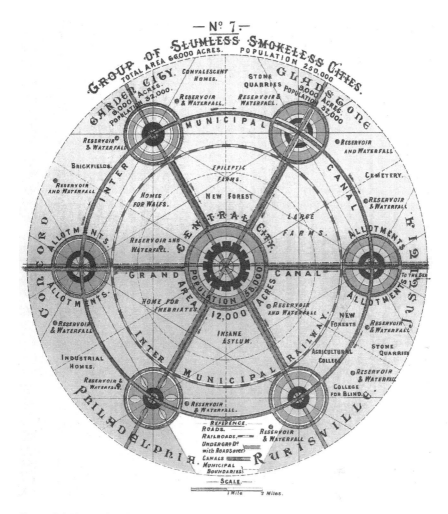

Figure 2.1 Howard's Diagram no. 7 (reproduced from Howard, 1898). In the original, all the land that is not shown as developed is coloured green.

and was shortly followed by others in England (Welwyn, 15 miles south of Letchworth, begun in 1920) and elsewhere (such as Hellerau, near Dresden in Germany, begun in 1903). Beyond the construction of specific towns, however, Howard's ideas, including the green belt, began to be developed in different ways by others, both in general terms and in relation to specific proposals, often in relation to London.

In 1901, Lord Meath, Chairman of the London County Council (LCC) Parks and Open Spaces Committee publicly requested the LCC and 'the other metropolitan authorities … to furnish London with a green girdle, linking the parks with one continuous chain of verdure' (cited in Loftus Hare, 1937, p. 678).

This girdle would be an oval ring, 5 to 8 miles from the centre of London (taken as Charing Cross) and incorporate 12,696 acres (5,078 hectares) of open spaces, including such famous areas of parkland as Alexandra Palace and Park, part of Epping Forest, Richmond Park and Blackheath. Others had similar ideas, among them D. B. Niven, who in 1909 'proposed a ring road and a depressed railway around London about ten miles from the center' (Foley, 1963, p. 15), and George Pepler, later President of the Town Planning Institute and Chief Planning Inspector to the UK Ministry of Health, who envisaged 'a great girdle around London' (Pepler, 1911, p. 614). The purpose of this would be not to stop the city growing but to facilitate concentric communication in order to reduce congestion in the city centre. It would also 'profitably conserve a belt of green, link up existing suburbs, and make provision for the future ... [and] open up a great deal of fresh land, which, if properly town-planned, could form an almost continuous garden suburb around London' (ibid., pp. 614–15). The aim of the latter would be to encourage the population to take exercise and improve their health – something also identified as important by Howard (see above) and another influential early town planner, Raymond Unwin, architect of Letchworth, who subsequently became technical advisor to the Greater London Regional Planning Committee (GLRPC).[2]

Unwin, like Howard before him, felt there was a need to limit the size of towns. He was inspired by ancient towns, enclosed as they often were by walls, which meant that they did not possess 'that irregular fringe of half-developed suburb and half-spoiled country which forms such a hideous and depressing girdle around modern growing towns' (Unwin, [1909] 1994, p. 154). While recognising that city walls were no longer required or justified, Unwin was keen to reinstate some form of limit to the size of towns and cities – if not in population terms, then in terms of the physical area they encompassed. He desired an 'intervening belt of park or agricultural land' (ibid., p. 154) between urban areas. These belts would, Unwin thought, have various purposes. They would be administrative boundaries, 'to foster a feeling of local unity ... breathing spaces ... haunts for birds and flowers, and as affording pleasant walks about the towns ... they ... would in a very true and right way bring into the town some of the charm of the country' (ibid., pp. 163–4). The latter phrase carries echoes of Howard's town–country magnet, perhaps deliberately. Indeed, others have noted the commonality between Howard and Unwin's conceptions of the green belt as a device to facilitate access to open recreation land for city dwellers, in the face of opposition from groups such as the CPRE, 'as they felt it would increase pressures on agriculture and the rural landscape' (Elson, 1986, p. 6).

Notwithstanding any opposition in the UK, green belts were exerting an influence internationally. For example, the '1924 Amsterdam International Planning Conference, where ... a highly influential model was that of the core city ringed by the greenbelt' (Sorensen, 2002, pp. 144–5), played a role in the development of, among others, the Japanese planning system and the plan for Greater Tokyo. In the USA, the town of Radburn in New Jersey was built to a plan produced in 1928–9. Radburn, which subsequently became very influential, 'inverted the greenbelt by placing the green areas *within* the residential areas' (Clapp, 1971, p. 35, emphasis added) – a very different approach, but with a common aim of ensuring access to green spaces for recreational purposes. This in itself was replicated on a much larger scale with the Dutch *Randstad/green heart* approach – see Chapter 5.

The story of the green belts begins to develop more quickly at this point, with the 1930s bringing the first attempts to create a large-scale green belt around London and increased public commentary on the negative impacts of urban sprawl in the UK.

The 1930s

As in the 1920s, the story of the green belt in the 1930s can be examined both in relation to the example of London and in general terms. Taking the latter first, three "Greenbelt Towns" were built in the USA as part of the *New Deal*. These were Greenbelt, Maryland; Greendale, Wisconsin; and Greenhills, Ohio. Green belts were integral to these towns and had some specific purposes: 'to provide a balance between urban and rural land use areas, a geographic limitation for the size and density of the community, and a wall of protection from external encroachment' (Clapp, 1971, p. 34). Parsons (1990) provides a useful summary of the development of the Greenbelt Towns, including the influence of planning in England and elsewhere.

In England, three men are often credited with devising and popularising the green belt notion: Ebenezer Howard, Raymond Unwin – whom we have heard about already and return to below – and Patrick Abercrombie. Abercrombie's work during and immediately after the Second World War will be discussed in the next section, but he was active much earlier, having become Lever Professor of Civic Design at the University of Liverpool in 1915 and moved to University College London as Professor of Town Planning in 1935. He was Honorary Secretary for the CPRE, then known as the Council for the Preservation of Rural England, from its formation in 1926, which perhaps explains his contributions to two books written in this period that we wish to examine here, illustrative as they are of a quantity of thinking and commentary going on at this time which emphasised the risk that urban sprawl posed to the *beauty* of the British countryside.

Both books were produced by Clough Williams-Ellis, the architect who created the mock-Italian village of Portmeirion in north Wales. The first book, published in 1928, was *England and the Octopus*, written by Williams-Ellis with an epilogue by Abercrombie. Williams-Ellis advocated a return to the values of 'our great- or great-great-grandfathers. They had a general sense of order and beauty' (Williams-Ellis, [1928] 1975, p. 12). Williams-Ellis felt this sense had been lost, to the detriment of the nation: 'it is chiefly the spate of mean building all over the country that is shrivelling up the old England – mean and perky little houses that surely none but mean and perky little souls should inhabit with satisfaction' (ibid., p. 15). He talked of the 'filching' of much of southern England by such development, going so far as to suggest that the enthusiasm with which men volunteered to fight in the First World War could be ascribed to the poor quality of the built environment and their consequent keenness to leave their homes (ibid., pp. 20–21). He was also concerned that many of those moving to live in the 'mean and perky little houses' he abhorred had 'no instinct for country life' (ibid., p. 38) – a concern shared by contributors to the second book, published in 1938. This was edited by Williams-Ellis, had a chapter contributed by Abercrombie, and was titled *Britain and the Beast* (perceptive readers may note a theme here).

Contributors to *Britain and the Beast* developed the idea of a fundamental difference between the rural and the urban – not merely in the character of the town vs. the countryside but, perhaps more profoundly, in the attitude of the people who lived in these different parts of England: 'the people are not as yet ready to take up their claim [to the countryside] without destroying that to which the claim is laid' (Joad, 1938, p. 64); 'The nightingale, we fancy, cannot endure the waste paper and cigarette cases which are the slot [*sic*] of the urban migrant' (Beach Thomas, 1938, p. 201). Contributors referred to town dwellers visiting, or seeking to live in, the countryside by such terms as 'invasion' (Kaye-Smith, 1938, p. 32), using what to us appears to be intensely patronising language to describe both urban in-migrants and the "indigenous" rural population. The former are described as 'those who cultivate the coarse and vulgar practice of always going anywhere' (Beach Thomas, 1938, p. 208), the latter as apparently simple-minded folk who needed to be protected. Summer visitors to the south-east coast, and their desire for boat trips,

> has meant easy money for the fishermen. And to honest folk easy money cannot be entirely good money – it is tainted … And the taint displeaseth God … The fishermen see the outward sign of this displeasure in the troubled heaving of the sea.
>
> (Lockley, 1938, p. 225)

There is of course little doubt that urban sprawl was causing problems, not just regarding the loss of beauty – Osborn (1946) expressed concern about the time and cost of commuting for the less well-off – but the contributions to *Britain and the Beast* appear most concerned about the visual impact of sprawl. The language used was quite extraordinary in some cases, perhaps reaching its apogee with 'The value of a W.C.[3] is vastly overrated when it is set above that of the aesthetic. An ugly house with a bath is less of an asset than a beautiful house without one' (Scott-Moncrieff, 1938, p. 270). It is of course easy to pour scorn on the language used by people writing nearly eighty years ago, but these extracts do illustrate an apparently not uncommon view held by some at the time. Patrick Abercrombie's contribution to *Britain and the Beast* is far more temperate than those we have quoted here, so we are not suggesting he shared all the views of other contributors, but he was evidently sympathetic to those of Williams-Ellis.

A concern with the aesthetic also featured in the policies of government at this time, leading to several attempts to limit urban sprawl. Reporting on the debate in Parliament around the Restriction of Ribbon Development Act of 1935, 'the

Parliamentary Secretary to the Ministry of Transport said "The whole of that four and a half miles of road between Dunstable and Luton is bordered with most ghastly little houses; behind is beautiful downland which is quite invisible"' (Laws, 1955, p. 265). The 1935 Act was preceded by the Town and Country Planning Act of 1932, which allowed local authorities to prevent development in certain parts of their jurisdiction by a form of zoning. However, if they did so they were obliged to compensate landowners at full market value, so few local authorities took the opportunities available to them. The inadequacy of these pieces of legislation led to the Green Belt (London and Home Counties) Act, 1938, the genesis of which was both the sort of commentary discussed above and the work of, among others, Raymond Unwin.

As noted above, Unwin was appointed as technical advisor to the GLRPC at the beginning of 1929. He saw it as vitally important to reserve 'an adequate proportion of open space … it is hoped that the provision … may take the form of a green girdle or chain of open spaces which would set limits to the solidly built up areas of London' (*The Times*, 12 August 1929, cited in Miller, 1989, p. 25). In his first report, published in January 1930, Unwin presented two ideal options for the development of London: the provision of small areas of green space (a 'green girdle') around and within built-up land; or the provision of a large, wide green belt with decentralised areas of building. Foley (1963) suggests Unwin preferred the former, but it is undoubtedly the latter which became more influential. Either way, political differences among the constituent local authorities, and a lack of funds, meant the proposals in his report were not implemented at that time. Unwin also presciently noted the problems caused by a lack of statutory regional planning (Miller, 1989), an observation which recurred in the wartime reports on planning (see the next section). Unwin's second report, published in 1933, proved more effective, as his more detailed proposals for a green girdle and the protection of playing fields became central to London County Council's Green Belt scheme, adopted in 1935 and given statutory weight by the 1938 Green Belt Act.

The 1935 scheme and the 1938 Act were both intended to preserve land in and around London from development through covenanting and land purchase. The latter was initiated by London County Council (LCC), which secured £2 million from the government to acquire land for public open space for Londoners. This was given as grants to adjoining local authorities to cover up to 50 per cent of the costs of acquiring land (Elson, 1986). The preamble to the Act states its purpose as 'the enhancement of the "health and amenities" of the people living near the Green Belt' (Mandelker, 1962, p. 28). However, Loftus Hare (1937, p. 677) felt that, even if they didn't use it, the green belt 'should bring indirect good to other Londoners, who may never see it, if it reduces the "sprawl" of London'.

For an excellent discussion of the detail behind the acquisition and reservation of land for green belt in and around London, see Amati and Yokohari (2006, 2007). Their fascinating work asks a number of important questions about this process and highlights conflicts between various actors, including the GLRPC and the LCC. They argue that, by the late 1930s, preserving (green belt) land had become all-important to some and 'took precedence over other pressing concerns such as rehousing London's poor' (Amati and Yokohari, 2007, p. 330).

Estimates vary of the amount of land reserved through the London green belt scheme – Elson (1986) puts it at 10,371 hectares by 1956, Amati and Yokohari (2006) at a much higher 28,600 hectares by 1939. Other cities had similar schemes, among them Birmingham, Oxford and Sheffield (Elson, 1986). As for the London councils, there were various objectives to the acquisition of such land for green belt, including stopping urban sprawl, protecting water supplies, and facilitating recreation (Ward, 2004; Elson, 1986). Ward (2004) notes that this confusion of purpose was inherited by the formal green belt which emerged following the Second World War.

The issues discussed so far, notably the loss of agricultural land for development and the concentration of the population in urban areas, principally London, were thrown into sharp relief by the difficulties in importing food during the Second World War and the Luftwaffe's bombing campaign, aimed principally at the industrial and military centres of the country, which were also the main population centres. This led to various commissions and committee reports during and immediately after the war and subsequent advisory plans for the future of London and other cities.

The war-time reports and advisory plans

Barlow, Scott, Uthwatt and Reith

Four reports (shown in Table 2.1) were commissioned during and immediately after the war to investigate various issues related to planning. Comprehensive summaries of the reports can be found elsewhere, but here we identify important discussion points or recommendations as they relate to green belts.

The Barlow report observes the problems with a lack of regional planning in England at the time and the specific problems posed by London attracting 'the best' (Barlow, 1940, p. 84). It concludes that the growth of London should be limited, with industry dispersed around the country. That was, however, predicated on an assumption that the population of Great Britain would not grow over the next forty years, and might even decline. This prediction was swiftly proven

Table 2.1 The "war-time" reports

Year	Chair	Title
1940	Sir M. Barlow	*Report [of the] Royal Commission on the Distribution of the Industrial Population*
1942	L. F. Scott	*Report of the Committee on Land Utilisation in Rural Areas*
1942	A. A. Uthwatt	*Final report [of the] Expert Committee on Compensation and Betterment*
1946	J. C. W. Reith	*Interim, Second Interim and Final Reports of the New Towns Committee*

false, as we will go on to discuss. The report is not wholly positive about garden cities, noting that 'it is doubtful whether the policy of creating either garden cities or trading estates is capable of very wide expansion' (ibid., p. 135). The CPRE's representative

> told the [Barlow] Commission that in general, controlled peripheral development was preferable to a Green Belt and satellite towns; the danger of the Green Belt was that development would leapfrog it and start on the other side, threatening wider tracts of the countryside than before.
>
> (Hall *et al.*, 1973b, p. 49)

But Patrick Abercrombie, Chair of the CPRE and a member of the Commission, was a firm advocate of green belts. His influence is perhaps visible in the report, which concluded that, where they are built, garden cities

> should be as far as possible complete units and placed well outside the existing town so as to enjoy advantages of air and space not possible at the centre or in the immediate suburbs, and they should be protected by a belt of open country so as to avoid eventual coalescence with the existing town.
>
> (Barlow, 1940, p. 133)

The Scott report, two years later, appeared to disagree with this notion of the green belt:

> Green belt is a townsman's expression which embodies a townsman's point of view and has come, unfortunately, to mean a belt of open land – of common, woods and fields – to be 'preserved' from building ... But open land cannot be 'preserved' and such a concept is false. We conceive the green belt to be a tract of ordinary country, of varying width, round a town ... in essence the green belt is just a tract of the countryside ... [it should be designated] with the object of securing the conservation of good farm land ... or, in other cases, of securing the reservation of woodland which can be both scenically attractive and economically productive.
>
> (Scott, 1942, pp. 71–2)

The Scott report is unusual in taking a *rural* perspective on green belts. Green belts have been criticised elsewhere for being an urban-driven policy, giving little consideration to the rural point of view (Elson, 1986), so Scott is a helpful counterpoint in that context. As we will see, however, the report was to have relatively little influence on policy as it developed.

The Uthwatt report, also published in 1942, took a pragmatic view of the pre-war development that had taken place in rural areas, arguing that 'houses, factories, roads, etc., to meet the requirements of a growing community must necessarily encroach on the rural areas. The defect lay in the fact that for the most part development was allowed to proceed unplanned' (Uthwatt, 1942, p. 12).

The report recommended 'the acquisition by the State of the development rights in undeveloped land' (ibid., p. 27), subject to compensation, determined on a national basis and averaged in order to avoid floating land value issues. This would facilitate, *inter alia*, 'the reservation of green belts and National Parks, the control over the expansion of existing towns and cities, [and] the establishment of satellite towns' (ibid., p. 28). The report made a general point about planning which is illuminating:

> Town and country planning is not an end in itself; it is the instrument by which to secure that the best use is made of the available land in the interests of the community as a whole. By nature it cannot be static. It must advance with the condition of society it is designed to serve.
>
> (Ibid., p. 12)

It might be interesting to reflect upon the green belts in this context and ask whether they have advanced with the condition of society. We return to this question later in the book.

Finally here we turn to the Reith reports on new towns. The New Towns Committee published three reports, two of them interim, in order that the members might have some influence over the New Towns Act of 1946, which was being written as they were sitting. They argued that a new town 'need not of itself destroy the beauty of the normal countryside. It may enhance it, and will bring more people within reach of enjoyment of it' (Reith, 1946, p. 12), and they set out some detailed recommendations for the form new towns should take. Among these was that they should 'be surrounded by open country and must not coalesce with other settlements' (ibid., p. 9). There should be two kinds of green belt: firstly, an agricultural belt of approximately three-quarters of a mile in depth, which should be owned and managed by the new towns agency in order to rearrange farm tenancies; secondly, a wider belt 'separating the town from other towns ... [which] should be safeguarded by normal planning control' (ibid., p. 13). Green space would also exist within the new towns, to be used for, *inter alia*, allotments.

As this brief discussion shows, these four reports, parts of which went on to form the basis for the landmark 1947 Town and Country Planning Act, contained divergent views on the form green belts should take, their scale and their purposes. As we will go on to see, it is fair to say that at no stage was there universal agreement on these issues, though, for a period of twenty years or so, one man's opinion carried a great deal of weight.

Abercrombie's advisory plans

During and shortly after the war, Patrick Abercrombie published, with various co-authors, plans for cities and towns such as Glasgow, Birmingham, Kingston upon Hull, Warwick, Bath and Plymouth. These plans were usually advisory, so their policies carried in general no statutory weight, but they were influential, to a greater or lesser extent, in the development of those and other cities.

Probably most influential was Abercrombie's plan for London, which actually comprised two documents, the 1943 *County of London Plan* and the more detailed 1944 *Greater London Plan*. These were in some ways flawed from the outset, as they were based, in common with the Barlow report, on the assumptions that the national population would not increase and that the population of Greater London would be reduced as a result of the relocation of industry through central state planning. As we know, the opposite in fact happened. This does not, however, invalidate Abercrombie's plans or reduce their interest to us. Hall *et al.* (1973b) argued that Abercrombie felt town and country to be opposite but complementary and that they should thus be kept apart. This notion imbues his plans.

Abercrombie started the *Greater London Plan* with a laudatory comment about the green belt secured through the 1938 Act: 'the [monies spent] are already paying handsome dividends in health and happiness, besides other probable returns in terms of money' (Abercrombie, 1945, p. 3). He aimed to designate as green belt in his plan most of the land acquired under that Act, and 'much more open land, not necessarily in public ownership but permanently safeguarded against building' (ibid., p. 8). Abercrombie endorsed the Scott report's view of the use of green belt for both agricultural and recreational purposes, the latter in the form of 'organised large-scale games' (ibid.) and the enjoyment of parks, woodlands and footpaths through farmland. He was clear, however, that agriculture was of primary importance: 'From the agricultural point of view, the preservation for farm and other cultivation use of the most productive land is essential ... the recreational use is to be made to interfere as little as possible with farming operations ...' (ibid., p. 11).

Abercrombie proposed four "rings" encircling the county of London. Proceeding outwards from the centre, these were the Inner Urban, Suburban, Green Belt and Outer Country Ring. In the last, Abercrombie foresaw the development of new towns, but only in appropriate locations, as determined by the relevant local authority: 'Certain local authorities have indicated to us, and we cordially endorse, the undesirability of *invading* particular districts by even the best planned new communities' (Abercrombie, 1945, p. 8, emphasis added; note the similarity with the language used in *England and the Octopus* and *Britain and the Beast*). He echoed the Reith report in seeing the need for smaller green belts around the new towns – and, indeed, around existing small towns beyond the main 'gigantic Green Belt round built-up London' (ibid., p. 11). He also advocated green wedges 'carried right into the centre of London' (ibid.).

The relationships between some of the key protagonists in the development of ideas about the green belt recur throughout this period. One chapter of the *Greater London Plan* was written by Dudley Stamp, vice-chair of the Scott committee. In that chapter, under the title "The Urban Fence", Stamp wrote: 'Of all the trends unfortunately characteristic of the years between the wars, the undignified sprawl of towns and suburbs into the surrounding country is one of the most to be deplored, from both the urban and rural points of view' (Stamp, 1945, p. 95). The "urban fence" was evidently conceived as an urban growth boundary, recalling the proposals of Unwin (1909) for a modern version of the ancient city wall.

Of Abercrombie's plans for other parts of the United Kingdom, Frey contrasts those for London and for Glasgow (the latter written in 1946). In London, the green belt was used as a 'means of containment' (Frey, 1999, p. 17), but, in Glasgow, Abercrombie conceived the green belt as a backdrop against which development took place: 'The outer line of the green belt follows the surrounding moor land edge and contains the Clydeside conurbation and all the open farming land surrounding it' (ibid., p. 18). In the Birmingham plan of 1948, Abercrombie combined decentralisation and peripheral development, with the green belt not necessarily being permanent (Elson, 1986). His 1943 Plymouth plan is well known for including the comprehensive redevelopment of a city centre badly affected by bombing, but this plan was also intended to be 'a regional plan for south-west Devon and south-east Cornwall' (Essex and Brayshay, 2005, p. 240) delivered through a voluntary joint regional planning committee. The regional elements of the Plymouth plan, including the redistribution of 40,000 people beyond the city's boundaries, failed because of profound disagreements with the surrounding authorities, not helped by Plymouth's parallel attempts to secure a boundary extension (see below for discussion of similar disagreements elsewhere later in the century).

As we noted above, Abercrombie's plans influenced planners in the UK and elsewhere. They were less effective in guiding the development of London in particular, in part because the assumptions of the era in terms of population growth were so rapidly outstripped. More fundamentally, however, other assumptions about the appetite of government for large-scale national and regional planning, including the relocation of industry and a substantial programme of new town building, were also proven false. Some ascribe a portion of blame to Abercrombie and his colleagues for this (Foley, 1963; Elson, 1986), with Foley in particular noting that, while the advisory plans were detailed in terms of the pattern of physical development they thought desirable, they did not consider in any depth 'the political economy in its broad operation nor the implementation measures and stages' (Foley, 1963, p. 172). The advisory plans have thus been criticised for glossing over the potential difficulties inherent in their implementation (Foley, 1963; Hall, 1974). Equally, however, we can blame the government, both at the time and later, for having the will to implement only one part of what Elson (1986, p. 47) called 'a triad of policies (Redevelopment – Green Belt – New and Expanded Towns)'. It seems obvious to note that, if one uses a green belt or other method to constrain the expansion of a town or city which has a growing population, but does not actively redevelop the centre or build sufficient satellite settlements, the end result will simply be the increase of urban density and land prices. As we will go on to discuss, this is indeed what happened in many places.

The immediate post-war years and the 1955 circular

The 1947 Town and Country Planning Act is widely regarded as setting the foundations for the planning system in the UK that is still in place today. Again, there are other sources with comprehensive reviews of its provisions and impacts (see, for example, Cullingworth *et al.*, 2015), but the three core elements are of relevance to us:

1   the nationalisation of development rights, with landowners forced to apply for planning permission to develop their land;
2   the payment of compensation for this nationalisation, at a fixed rate from a national pot of £300 million (equivalent to approximately £11 billion today) – there was a corresponding development charge, whereby the increase in land value as a result of planning permission was nationalised;
3   the requirement for local authorities to prepare development plans with a twenty-year lifespan.

The compensation was to be paid by the central government, not local authorities, as was previously the case under the 1932 Act. This removed one major impediment to local authorities creating green belts. However, they were not obliged, or even encouraged, to do so, and consequently few did. The 1951 Conservative government was elected on (among other things) an anti-planning platform. It left most of the Act in place but abolished the development charge. But the concept did make two further appearances, in 1967–71 and 1976–9 (again, see Cullingworth *et al.* (2015) for more detail). There was a large rise in the population between 1951 and 1961, something that had not been predicted by Abercrombie or the architects of the 1947 Act. The "baby boom" meant that the careful planning of new towns was not enough, and restrictions on private development were eased to meet the need/demand for housing. The development control system, which had been designed to deal with only a small amount of private development, couldn't cope. Land allocations were exhausted, and land became a valuable commodity (Hall *et al.*, 1973a).

Public (and increasingly private-sector) development of housing and other uses continued at a high rate in the early 1950s, and some of this was (proposed) on previously designated green belt. Beyond London there seemed to be little appetite among local authorities to stop peripheral expansion (Elson, 1986). The Town and Country Planning Association was one organisation lobbying for a limit to such development (Laws, 1955), as was the architectural commentator Ian Nairn, who that same year published the polemic *Outrage* (Nairn, 1955), in which he coined the term *suptopia* to describe urban sprawl. Perhaps partly in response to this lobbying, the Minister of Housing and Local Government, Duncan Sandys, made a statement in April 1955 in the House of Commons, in which he said: 'I am convinced that, for the well-being of our people and for the preservation of the countryside, we have a clear duty to do all we can to prevent the further unrestricted sprawl of the great cities' (MHLG, 1956, p. 55). Sandys, in a speech at Welwyn Garden City on 30 April 1955, further clarified his intended extent of green belts: 'he said that he wanted to see plans for green belts not only around the big cities but around all towns … It was necessary to define the limits of towns, and to make it "damned difficult" to go past the limits' (Town and Country Planning Association, 1955, p. 277). These speeches were swiftly followed, in August 1955, by MHLG Circular 42/55, which formalised the request to local authorities to establish green belts. Given the importance of this circular for green belt policy for the next thirty years, we reproduce it in full in Box 2.1.

Box 2.1   Circular 42/55

1) Following upon his statement in the House of Commons on April 26th last (copy attached), I am directed by the Minister of Housing and Local Government to draw your attention to the importance of checking the unrestricted sprawl of the built-up areas, and of safeguarding the surrounding countryside against further encroachment.

2) He is satisfied that the only really effective way to achieve this object is by the formal designation of clearly defined Green Belts around the areas concerned.

3) The Minister accordingly recommends Planning Authorities to consider establishing a Green belt wherever this is desirable in order:

   a) to check the further growth of a large built-up area;
   b) to prevent neighbouring towns from merging into one another; or
   c) to preserve the special character of a town.

4) Wherever practicable, a Green Belt should be several miles wide, so as to ensure an appreciable rural zone all round the built-up area concerned.

5) Inside a Green Belt, approval should not be given, except in very special circumstances, for the construction of new buildings or for the change of use of existing buildings for purposes other than agriculture, sport, cemeteries, institutions standing in extensive grounds, or other uses appropriate to a rural area.

6) Apart from a strictly limited amount of 'infilling' or 'rounding off' … existing towns and villages inside a Green Belt should not be allowed to expand further. Even within the urban areas thus defined, every effort should be made to prevent further building for industrial or commercial purposes; since this, if allowed, would lead to a demand for more labour, which in turn would create a need for the development of additional land for housing.

7) A Planning Authority which wishes to establish a Green Belt in its area should, after consulting any neighbouring Planning Authority affected, submit to the Minister, as soon as possible, a Sketch Plan, indicating the approximate boundaries of the proposed Belt. Before officially submitting their plans, authorities may find it helpful to discuss them informally with this Ministry either through its regional representative or in Whitehall.

8) In due course, a detailed survey will be needed to define precisely the inner and outer boundaries of the Green Belt, as well as the boundaries of towns and villages within it. Thereafter, these particulars will have to be incorporated as amendments in the Development Plan.

9) This procedure may take some time to complete. Meanwhile, it is desirable to prevent any further deterioration in the position. The Minister, therefore, asks that, where a Planning Authority has submitted a Sketch Plan for a Green Belt, it should forthwith apply provisionally, in the area proposed, the arrangements outlined in paragraphs 5 and 6 above.

I am, Sir,

Your obedient Servant,
A. B. VALENTINE

(MHLG, 1955, pp. 1–2)

Several parts of the circular are of importance, and the listing at paragraphs 5 and 6 of the constraints on development within a green belt are clearly central to the effectiveness of the designation. These have remained largely unchanged. The aim and objectives of the green belt specified in paragraphs 1 and 3 are also interesting (government would approve green belts only if they met one or more of these), and they remained unchanged until 1984, though they were subsequently added to incrementally. Finally, the note at paragraph 9 was truer than A. B. Valentine could have known: we are still, more than sixty years later, awaiting comprehensive green belt boundaries in local plans!

The circular seems to have been in part a political initiative. Mandelker (1962, p. 30) observed that Sandys issued it 'over objections from his technical staff', and Elson (1986, p. 15) noted criticisms of the concept by the ministry's Permanent Secretary, who was concerned that the question of where development was to go, if not in green belts, had not been answered. Others were enthusiastic about it: the President of the Town Planning Institute lauded it in his presidential address in 1955 (Elson, 1986; Amati, 2008), the Town and Country Planning Association (TCPA) embraced the concept, and rural local authorities were very happy to have a tool with which to limit the physical and administrative expansion of urban areas (Mandelker, 1962; Hall, 1974; Elson, 1986). This last role, as a "stopper" of urban growth, meant that the green belts were fundamentally different from the concept put forward by Ebenezer Howard (Elson, 1986). There was no mention in the circular of the positive potential of green belts in terms of agriculture or recreation – an omission that may have 'reflected the prevailing assumption of the period that the recreational needs of town dwellers would largely be met by the establishment of National Parks and the provision of access to other areas of the countryside' (Murray, 1991, p. 54). However, the TCPA was keen to obtain clarification of this and other issues, so wrote to Sandys. His reply (beginning 'My Dear Osborn')[4] was published in *Town & Country Planning*. In the following extract, the TCPA's comments are in Roman text and Sandys's responses in italics:

The primary purpose of a green belt is to set a limit to urban expansion and help to preserve rural areas for the enjoyment of town dwellers. *Agreed* … Agriculture should be the primary land use in green belts. *Agreed* … In determining the size of green belts the main consideration should be that it should be wide enough to travel to work across it. The memorandum suggests … a minimum of five miles and maximum of ten miles … *The Minister has preferred not to prescribe any uniform maximum or minimum widths … He thinks, however, that the figures suggested by the Association are useful to local planning authorities as a rough guide* … For large conurbations the green belts will be at some distance from the centre. Therefore it is important to preserve green wedges. *Agreed*.

(Sandys, 1956, pp. 151–2)

This correspondence, which of course carried no weight in planning decisions, allows us an insight into what the originator of the first formal green belt planning guidance expected. These early ideas had little influence over how green belts evolved, as we shall see.

Local authorities were quick to respond to the circular. In its 1956 annual report, the MHLG noted the receipt of various sketch plans but was evidently keen to limit these.

The establishment of a green belt is, however, a step which calls for most careful deliberation. The strict control of development that needs to be exercised in green belt areas in order to resist the pressure of urban sprawl cannot be justified except where the position urgently demands it [re the three justifications in the circular].

(MHLG, 1957b, p. 43)

The ministry was also keen to make clear that green belt boundaries should be carefully defined to avoid the need for amendment. To assist in the latter, a second *Green Belts* circular (50/57) was issued in 1957, with detailed guidance on boundary definition. The inner boundary would 'mark a long-term boundary for development' (MHLG, 1957a, p. 1). To ensure that sufficient land was made available for development, Circular 50/57 introduced the concept of "white land" – land that might be needed in future development plans. Elson (1986) reviewed the sixty-nine sketch plans that were submitted between August 1955 and December 1960. Some areas of green belt were approved, some trimmed and some rejected. Elson concluded that the decisions on these plans formalised the concept of the green belt as a "stopping" device. As noted above, this was not the rounded definition of green belts that Ebenezer Howard, among others, imagined. It was also a problem in the context of housing shortages and the consequent increases in land price in the 1960s (Mandelker, 1962; Cullingworth, 1980).

The 1960s and 1970s

It was partly in response to those problems that, in 1960, Circular 37/60 (*The Review of Development Plans*) was issued, noting the government's desire to 'see more land provided for development (where that did not conflict with important planning objectives)' (MHLG, 1960, p. 1) and the need to match the restriction on development imposed by green belts with 'adequate provision for balanced and compact development elsewhere' (ibid., p. 2). Such development should be both within and, crucially, *beyond* the green belt. This was not straightforward, and local authorities beyond green belts were typically not enthusiastic, since the financial incentives on offer were inadequate to overcome concerns over large numbers of incomers. Similarly, outside London and the South East, urban authorities were reluctant to reduce their tax base by dispersing their population beyond green belts into other local authorities.

Local plans were not meeting the forecast need for housing, with a 2 million shortfall between population projections for 1961–81 and the number of houses planned for. Speculative developers therefore sought additional sites and focused their attention on the edges of green belt towns, where the difference in value between land with and without planning permission was highest (Elson, 1986). The government concluded that more land should be made available for development to reduce land prices. MHLG officials carried out studies of each region and stopped approving green belt boundaries in 1962.

Also in 1962, the MHLG published a booklet titled simply *The Green Belts*. The booklet, apparently designed for public consumption and not to indicate any change in policy, is nonetheless useful to provide an insight into wider government thinking on the topic at the time. It specifies a purpose, reasons and object behind green belts, which are also instructive.

> The primary purpose of a green belt [is] ... to prevent ... sprawl ... The secondary purpose is perhaps better understood and appreciated. It is to provide the townsman with the opportunity to escape from the noise, congestion and strain of city life and to seek recreation in the countryside.
>
> (MHLG, 1962, p. 1)

This additional *purpose* fills the gap noted above in Circular 42/55. The *reasons* for establishing a green belt are as set out in the circular (see Box 2.1), and the booklet adds an *object* of including land in a green belt – to 'keep it permanently open' (MLHG, 1962, p. 7). The terms purpose, reason and object(ive) all have different meanings, but to us these differences are subtle, and the use of the terms in the booklet may indicate an attempt to broaden the justification for green belts to shore up public support for them. It goes on to emphasise the importance of matching green belt restraint with decentralisation (i.e., the development of new towns), which, as noted above, was not proceeding at any great pace.

The problems in London and the South East were particularly acute. Cullingworth (1980, p. 235) analysed government papers, including some from the Cabinet and the Cabinet Committee on Population and Employment, and found that, in drafting the 1963 White Paper *London: Employment, Housing, Land*, the government concluded that it was essential to build housing on some green belt land. Keith Joseph, Minister of Housing and Local Government, was convinced that he could not 'dodge' this. Indeed, 'I think it is time that I started to clear the air. Nobody intends to destroy the green belt. On the other hand, the pressure for housing land within easy reach of London is tremendous.'

(Cited in Cullingworth, 1980, p. 235)

The idea was to reallocate some green belt land for housing but subsequently extend the green belt in other areas. A pragmatic approach to green belts appears to have had cross-party support at this time. Richard Crossman, the first Labour Minister of Housing and Local Government in the 1964–70 Labour government, gave permission for three green belt deletions. Ward (2004) quotes his diary from December 1964:

'I've decided, if rigidly interpreted, a green belt can be the strangulation of a city. With so many people to house we can't put them all in New Towns thirty miles away on the other side of the green belt … in some cases we shall trespass into the green belts or turn a green belt into four or five green fingers.'

(Cited in Ward, 2004, p. 152)[5]

The problem came in the divergence between the private thoughts of ministers and their public pronouncements, which were usually strongly pro-green belts. This, coupled with strong public support for green belts, meant local authorities were intensely reluctant to endorse urban extensions (Elson, 1986). The 1964–70 Labour government attempted to circumvent this by introducing advisory regional economic planning through *regional studies*. The first such study, of the South East, was published in that year and proposed new and expanded towns to accommodate 1.25 million people by 1981, one of which was Milton Keynes. Extensions to the green belt should not proceed as proposed in draft local plans, but the original green belt should be retained. The study also discussed but rejected two ideas – pushing back the green belt boundary by 1 kilometre all around and turning the London green belt into a series of wedges. Further regional studies of the South East in 1967 and 1970 proposed a more positive approach to green belts, allowing some development (this was rejected) and growth beyond the green belt. Elson saw this as 'a victory for the Home Counties' (Elson, 1986, p. 36), which vigorously resisted changes to the approved green belt and tightened their green belt maps, "washing" green belt over many small settlements and thus ruling them out as locations for development.

Similar controversy raged elsewhere in England, the West Midlands being another example of conflict between a central urban authority (Birmingham) and surrounding rural authorities, including Worcestershire, which contested attempts by the city to expand its boundary. Industry was likewise unwilling to relocate beyond the green belt, which 'bottled up such forces in the conurbation that the green belt had to give way' (Elson, 1986, p. 37). The latter was achieved only by the government approving planning applications on appeal on the edge of Birmingham City (see Chapter 1 for details of the English system of planning "by appeal").

At around this time there was a degree of criticism of the green belt concept, and its utility in the context of the time, from such diverse voices as Dame Evelyn

Sharp, former Permanent Secretary to the MHLG (Sharp, 1969) and Peter Hall (Hall *et al.*, 1973b; Hall, 1974). Hall criticised the green belts and the approach of 'containing' urban England on a number of fronts, including questioning their regressive effects. We will return to these arguments in Chapter 4, but Hall also noted the ineffectiveness of the green belts in supporting recreation for town dwellers. He observed how little of the green belt was in recreational use (5.4 per cent of the London green belt, by his calculation) and opined that, although much of the rest of it was accessible via public rights of way, the average 'townsman' was unlikely to use it, because they did not know how to use an Ordnance Survey map. While this teeters on the edge of the sort of language cited in the section on the 1930s above, not least his description of 'a litter of children in the back of the car' (Hall, 1974, p. 6), Hall was absolutely right to point out that the aim of the early green belt pioneers (Howard principal among them) to provide green space for recreation and health had not been achieved. The 1968 Countryside Act allowed local authorities to create country parks, but these, on account of the traffic generated by people travelling to them, attracted opposition from local residents. Hall concluded that the protection of green belts at that time 'seems to be less a matter of planning policy than planning politics' (ibid., p. 8).

Another political issue at the time was local government reorganisation. The Redcliffe-Maud Commission (1966–9) was set up by the Labour government of the time to review the topic. It recommended a mix of *unitary* and two-tier local authorities, the former in some areas of intense green belt debate. Cheshire and Staffordshire County Councils, for instance, could have been abolished under this approach,[6] so the report was viewed as favouring urban areas. In the general election of 1970 the Conservatives were returned to power and abandoned these proposals in favour of a comprehensive two-tier system of counties and districts, which were eventually legislated through the Local Government Act of 1972, along with a division of planning responsibilities between counties (the structure plan) and districts (the local plan). The 1972 Act favoured rural areas in several ways. Metropolitan county boundaries were tightly drawn, limiting the potential both for urban authorities to expand and for green belt release. Further, before the Local Government Planning and Land Act of 1980, county councils had the power to direct the refusal of some planning applications contrary to the structure plan – for example, those that proposed green belt deletion (Murray, 1991).

So, as the 1970s began, the green belt, despite the criticisms noted above, seemed as secure as it had ever been as a result of the strong political support it enjoyed, along with the decline in pressures from growth in the late 1960s (Elson, 1986). But, by 1972–3, the economy had recovered somewhat and the size of households was falling, meaning that more housing land was needed (Ward, 2004). Circular 122/73, *Land Availability for Housing*, changed the status of "white land" – introduced as a concept in 1957 as land that may be needed in future development plans but was not to be developed in the current plan period. It made clear that, if a local authority did not have an up-to-date plan in place, white land could be 'regarded as suitable for housing' (DoE, 1973, Annex A, para. 5) if its development would extend existing development and be suitable in other ways. The circular did, though, maintain the status of the green belt, and Hall (1974) noted that this would mean there had been no change in the general approach to planning for more than twenty years. This was despite the far-reaching changes in society in that time, not least the rise in car ownership – the proportion of households in Great Britain with no car had fallen from 85 per cent to 45 per cent between 1951 and 1972. Hall felt that the two purposes of green belts spelt out in the 1962 *Green Belts* booklet – preventing sprawl and encouraging recreation – were contradictory, and this contradiction had never satisfactorily been resolved. Others expressed similar views, Jackson (1972) arguing that little consideration had been given to how to resolve this discrepancy in the very early days of green belts, and Elson (1986, p. 48) feeling that the whole idea of a dual-purpose green belt could be seen as a potential weakness to be exploited by developers, who 'could argue that land should be developed if no demand or accepted use for recreation existed'. The West Midlands green belt, in and around Birmingham, was 'most obviously defined with recreation needs in mind' (ibid., p. 60), with green wedges penetrating into the built-up area. Other green belts gave less consideration to recreation.

The criticisms of Hall, Jackson and others had little effect on public opinion, however, and the potential of using a green belt for recreation, even if this was never taken up, was apparently enough to ameliorate any objections on social justice grounds. Indeed, the amount of green belt proposed under the new system of local and structure plans introduced in 1968 increased the amount of approved green belt from 6,930 square kilometres to 15,815 square kilometres, 10.5 per cent of the land in England and Wales (Elson, 1986). Elson argued that this expansion occurred at the same time as outer-city areas, including green belts, came under increased development pressures for both economic development and housing. Government advice encouraged local authorities to be more lenient with regard to the location of industry and to have a five-year supply of housing land, but there was a continued emphasis on the importance of protecting green belts and agricultural land. Local authorities were left to make their own sense of these arguably contradictory policy directions.

1979–1990

The election of the Conservative government led by Prime Minister Margaret Thatcher in 1979 began a decade that changed many parts of British life. Neoliberalism and what has been called "the hollowing out of the state" had profound effects on the economy and society. These changes clearly affected the planning system, but during this time 'in no sense was it dismantled, or even changed in any significant way' (Cullingworth and Nadin, 2006, p. 28). The same was true of the green belts. They have been described as 'the exception to Thatcherism' (Ward, 2004, p. 226), as the amount of green belt approved increased during the 1980s. That statement, however, conceals a great deal of complexity and controversy around green belts and development in the countryside more broadly that ensued in the 1980s.

The fundamental contradiction for that Conservative government – and indeed for subsequent Conservative administrations – was how to rationalise the preferences of their core voter base in the (rural) South East, who tend to oppose development in the countryside, particularly on the green belt, with the neoliberal ideological preference to encourage and facilitate development by the private sector, which would prefer to build on exactly those areas (Elson, 1986; Ward, 2004). This contradiction came to a head twice during the 1980s, when two Secretaries of State for the Environment attempted, and failed, to reform green belt policy.

The first was Patrick Jenkin, SoS from 1983 to 1985. The House Builders Federation (HBF) was increasingly playing a role in the examination in public of structure plans, objecting to draft policies such as the 'outrageous attempts by Hertfordshire, Buckinghamshire and Surrey, Berkshire and Essex to define their entire counties as green belt' (Humber, 1982, p. 2). The HBF called for a new green belt circular that would return to what it saw as the original purposes of the green belt, instead of what it had become – 'an all-purpose means of preventing development over enormous areas of the country' (ibid.). Partly in response to lobbying of this type, Patrick Jenkin issued draft green belt and housing land circulars. These were carefully written to try and avoid controversy – 'An innocent reader ... might easily have assumed they were offering unswerving support for the green belt' (Ward, 2004, p. 227). But they were in fact suggesting that local authorities should delete small areas of green belt and replace them with land elsewhere. The CPRE and others mounted a campaign to protect the green belt, leading to a hearing of the House of Commons Select Committee, which recommended, rather than deletion, increased protection of green belts, including such smaller pockets of land (Elson, 1986).

The publicity campaign and select committee recommendations meant that the final green belts circular, issued in 1984 as Department of the Environment Circular 14/84, reiterated the government's commitment to green belts and removed any reference to deletion. It reconfigured the *role* (another new descriptive, to add to *purposes*, *reasons* and *objects*) of green belts as 'checking the unrestricted sprawl of built-up areas, safeguarding the surrounding countryside from further encroachment, and assisting in urban regeneration' (DoE, 1984, p. 1) and reaffirmed the objectives of green belts as those laid out in Circular 42/55 (see Box 2.1). The new circular emphasised that green belts should be protected in the very long term – 'as far as can be seen ahead' (ibid.), but that, where local authorities had not yet defined boundaries, these should be loose enough to allow space for development. Elson (1986) sees this approach as being an attempt by central government to gain more control over land release. With looser inner boundaries to green belts, inspectors and the minister could grant planning permission, on appeal, for housing on the periphery of urban areas without having to delete green belt. Conversely, large and extensive green belts strengthen the hand of local authorities which might wish to resist such peripheral development.

A second attempt by a Conservative SoS for the Environment to release more land in the countryside for housing was made by Nicholas Ridley, in post from 1987 to 1989. Rather than trying to amend green belt policy and continuing to support piecemeal urban sprawl, Ridley sought to promote, in his own words, 'a few new small towns and villages' (Ridley, 1991, p. 116), to be built by the private sector. Perhaps unsurprisingly, there was considerable opposition to these new settlements. One was on green belt land, so was refused, and Ridley approved one, in accordance with the report of a planning inspector. Shortly after, in 1989, he was replaced as Environment Secretary by Chris Patten, who reversed Ridley's decision and refused the subsequent two appeals, ending the short-lived policy.

In 1988 the government published *Planning Policy Guidance 2: Green Belts* (PPG2) and updated the 1962 *Green Belts* booklet. These two documents effectively combined one of the *purposes* from the 1962 booklet with the *reasons* from Circular 42/55 and added, from the 1984 circular, a *role* for green belts.

> Green Belts have five purposes:
>
> - to check the unrestricted sprawl of large built-up areas;
> - to safeguard the surrounding countryside from further encroachment;
> - to prevent neighbouring towns from merging into one another;
> - to preserve the special character of historic towns; and
> - to assist in urban regeneration.
>
> Green Belts also have a positive role in providing access to open countryside for the urban population ... Green Belts often contain areas of attractive landscape, but the quality of the rural landscape is not a material factor in their designation or in their continued protection.
>
> (DoE, 1988b, paras 4–6)

The 1988 booklet was published to update the 1962 booklet

> not because our Green Belt policy has changed but because it demonstrates the continuity of that policy and our strong commitment to the Green Belts ... The basic Green Belt policy and its original purposes have not changed. But in the 1962 booklet there was a lot of emphasis on 'decentralisation' out of London and the other conurbations to New Towns ... We now put much more emphasis on the **regeneration** of the older urban areas and on the **re-use** of urban land.
>
> (DoE, 1988a, p. 6)

Despite the claim of lack of change, other amendments to the green belt policy were made with the 1988 PPG2. The newly introduced purpose of safeguarding the surrounding countryside from further encroachment 'was the first reference to rural objectives for green belts' (DoE, 1993, p. 7). The authors of that report for the Department of the Environment also saw the altering of 'to check *further* growth' (MHLG, 1955) to 'to check *unrestricted* growth' to be significant in disassociating green belts from stopping growth. The fourth purpose was also amended, with 'historic' added to the guidance to deter non-historic smaller towns from instigating green belts (DoE, 1993).

Michael Heseltine, then Environment Secretary, announced in 1991 that he wanted green belts to be more positive. He suggested two new objectives: 'to increase opportunities for quiet enjoyment of the countryside; and to enhance and improve the natural beauty of the countryside adjoining towns and cities' (cited in DoE, 1993, p. 68). These new objectives were not adopted at the time, but versions of them came in in 1995 (see below). What did happen around this period was the approval of some large economic development projects in the green belt, including the National Exhibition Centre in Birmingham and the Nissan factory near Sunderland (Ward, 2004). Those were schemes to provide jobs in areas of economic decline, something that was evidently politically easier to approve than green belt deletions for housing.

The late 1980s and early 1990s were a time when the idea of the green belt as a "stopper" was receiving additional support from the green movement, which 'allowed an aura of spurious radicalism to surround the contented society of the outer cities, defending not just their backyards but the altogether grander notion of "the environment" from the incursions of the builder' (Ward, 2004, p. 229). "The environment" had risen up both the public and the political agenda during the 1980s, both internationally and domestically. In 1987 the report *Our Common Future* by the United Nations World Commission on Environment and Development (the Brundtland Commission) was published. In 1988 and 1989 Margaret Thatcher gave speeches to the Royal Society and the United Nations General Assembly, respectively, in which she firstly linked the protection of the environment with continued economic development (Thatcher, 1988) and secondly urged international action to address environmental damage caused by man (Thatcher, 1989).

The 1990s

This apparent shift in policy continued into the 1990s. In September 1990 the UK government published a White Paper, *This Common Inheritance: Britain's Environmental Strategy*, a title which reflected that of the earlier "Brundtland" report (WCED, 1987). The White Paper, though described as 'the most comprehensive document produced by a UK government on the environment' (Tromans, 1991, p. 168), also received a great deal of criticism both in the press and from green groups for being insufficiently ambitious. In November of 1990 Margaret Thatcher was replaced by John Major as Prime Minister. Neither the White Paper nor this change in leadership heralded any change in green belt policy, but they are important elements of the context within which that policy operated, as are two legislative changes, the 1990 Town and Country Planning Act and the 1991 Planning and Compensation Act. These two Acts, among many other things, introduced the notion of a *plan-led system*, under which local authorities are statutorily obliged to make decisions on planning applications in line with the development plan, unless 'material considerations indicate otherwise'. Comprehensive development plans were also made mandatory by the 1991 Act. These modifications to the planning system, introduced as amendments to the 1991 Act sponsored by the CPRE, have been described as a 'victory' for them and other environmental groups (Pennington, 2000, p. 73), reducing as they did the scope for developers to appeal refusals of planning applications. This was another example of the influence the CPRE was able to exert over the planning system. It has been argued that the agenda of the CPRE set the context for the 1947 Act and many subsequent iterations of planning law (Lowe and Murdoch, 2003), and we noted above the influence of Patrick Abercrombie, as Chair of the CPRE at the time, on the Barlow Commission.

In 1995 PPG2 was reissued, with the five purposes of green belts specified in 1988 left unchanged. The new version did, though, claim that it gave policy 'a more positive thrust by specifying for the first time objectives for the use of land in Green Belts' (DoE, 1995, preamble). Those objectives were:

- to provide opportunities for access to the open countryside for the urban population;
- to provide opportunities for outdoor sport and outdoor recreation near urban areas;
- to retain attractive landscapes, and enhance landscapes, near to where people live;
- to improve damaged and derelict land around towns;
- to secure nature conservation interest; and
- to retain land in agricultural, forestry and related uses.

(Ibid., para. 1.6)

The guidance stressed, however, that whether or not land fulfilled these *objectives* was irrelevant to designating or retaining it as green belt, with the *purposes* of green belts outweighing these objectives as planning considerations. The use of green belt as a stopper of urban growth, therefore, remained para-mount, and the area of approved land in England increased from 15,485 square kilometres to 16,523 square kilometres between 1989 and 1997 (Ward, 2004). This conception survived the change of government in 1997, with the New Labour administration maintaining, and arguably increasing, the focus on urban regeneration as a method to limit urban sprawl (Gunn, 2007). This is perhaps exemplified in the 1999 report of the Urban Task Force, commissioned in 1998 to examine how the then projected 4 million additional new households by 2016 could be accommodated in the UK's cities. The Urban Task Force made a series of recommendations, including a 60 per cent target for new housing to be built on previously developed land, sometimes known as brownfield land. That, and many other of these recommendations, was adopted by the government in its *Planning Policy Guidance Note 3: Housing* (DETR, 2000). Again, the CPRE was influen-tial here, since its Assistant Director Tony Burton was a member of the Urban Task Force, and the organisation had a major input into the document. Indeed, a CPRE policy officer claimed: "'We invented all the key planks in PPG 3. PPG 3 is basically CPRE policy'" (Murdoch and Lowe, 2003, p. 327).

The twenty-first century

At the turn of the millennium, then, the influence played by the CPRE in urban policy in the UK was as strong as it had been in 1947. This was one factor in the enduring strength of green belts, though strong public support for them made any suggestion of a change in government policy politically unacceptable (Ward, 2004). Indeed, apart from a small amendment to PPG2 in 2001 to refer to park and ride schemes, the government's planning policy on green belts in England remained unchanged until 2012. This was despite an increasing acknowledgement that the country was facing a 'housing crisis'. In 2003 the economist Kate Barker, a former member of the Bank of England's Monetary Policy Committee, was appointed by the government to undertake a study of the UK housing market. Her report recommended, *inter alia*, that, although the general principle of using the green belt to limit urban sprawl should remain unchanged, 'planning authorities should show greater flexibility in using their existing powers to change greenbelt designations where there are strong pressure points in a particular urban area, and where forcing development elsewhere would lead to perverse environmental impacts' (Barker, 2004, p. 44). The government's response to the Barker review stressed its agreement with the first part of this recommendation and ignored the second part, instead requiring local authorities to refer planning applications on green belt land to the SoS to introduce additional scrutiny to such proposals (HM Treasury and Office for the Deputy Prime Minister, 2005).

Despite this official policy position, there were some green belt deletions during the early 2000s, including around Newcastle, to facilitate the Great Park mixed-use development (see Figure 2.2), and in other parts of the country (Ward, 2004). These relatively small deletions, however, had only a marginal effect on the overall amount of designated land. Gunn (2007) undertook a review of all the regional spatial strategies (RSSs) under preparation at that time and found that several of them, notably in the East and South West of England, sought to alter their green belts in order to achieve 'more sustainable patterns of development … [by]

Figure 2.2 The Newcastle Great Park development (source: authors' own)

releasing some inner-edge green belt to allow urban extensions within the belt … It is argued that a looser green belt will enable towns and cities to develop in a compact way, as well as reducing leapfrogging' (ibid., pp. 610–11). This might make sense from an objective perspective but would doubtless have been less positively received by many of the general public – 84 per cent of whom, as noted in Chapter 1, believed in 2005 that the green belt should not be built upon. This in turn meant political support for green belts remained strong, and the three main political parties in the UK included commitments to protecting the green belt in their manifestos for the 2010 election (Conservative Party, 2010; Labour Party, 2010; Liberal Democrat Party, 2010).

The Conservative–Liberal Democrat coalition reaffirmed its support for green belts in its *Programme for Government* (HM Government, 2010) and introduced various reforms to the planning system during its five-year term in office. Many of these were justified under the badge of "localism", including abolishing RSSs, which featured "top-down" housing targets that local authorities were mandated to deliver. These reforms were in part intended to boost local support for housing, the assumption being that widespread opposition to development might be ameliorated if top-down planning was replaced by bottom-up, community-centred planning. While this idea was greeted with considerable scepticism, there is some evidence that, in some circumstances, local communities might welcome small-scale housing development if they have more of a say in its planning (Sturzaker, 2011; Sturzaker and Shaw, 2015). Ministers were clear, however, that they did not envisage the changes to the planning system resulting in any loss of green belt land – the reverse, in fact (Pickles, 2012), with the abandonment of most of the proposed strategic-level green belt deletions discussed in the preceding paragraph.

The National Planning Policy Framework (NPPF), published in 2012, combined all previous government planning policy, including PPG2. It retained the five *purposes* of green belts that had been in place since 1988 but removed any reference to the *objectives* of green belt land, simply stating that 'The fundamental *aim* of Green Belt policy is to prevent urban sprawl by keeping land permanently open' (DCLG, 2012, p. 19, emphasis added). This maintained the previous priority given to protecting green belts, but, as discussed in Chapter 1, another part of the NPPF was seen by some as increasing the pressure on local authorities to release green belt land. The NPPF strengthened the requirement for local authorities to identify a five-year supply of housing land that was considered viable for development. If a local authority could not identify five years' worth of housing land, its policies in relation to housing should be considered out of date, which meant that there would be a presumption in favour of sustainable development – i.e., local authorities should approve planning applications for housing in that situation unless they could demonstrate that the harm from doing so significantly outweighed the benefits.

Although the government continued to stress the primacy of the green belt, others, notably the CPRE and the National Trust, began to gather evidence that showed what they saw as the risk to the green belt. The CPRE, in August 2013, published analysis it had undertaken of emerging local plans, showing that 150,000 new houses were now proposed in the green belt (CPRE, 2013). Somewhat more quietly, it acknowledged that this was only 3,000 more than were proposed in draft regional plans before their abolition, perhaps undermining their claims of a substantial increase in green belt release, but equally challenging the government's assertion that they had increased protection for green belts. The National Trust followed this up in December 2013 with a survey of English local authorities, which found that half of those with green belt land 'are preparing to allocate some of it for development' (LGiU, 2013).

However, the rate of success at appeal on housing applications on green belt land remained low, with the SoS said to be reluctant to grant appeals 'ahead of any [green belt] release in the context of local plan making' (Dunton, 2014, p. 8). Dunton's article referred to an appeal decision to allow an application to double the size of Pinewood film studios in London's green belt – as in the 1980s, it was evidently an easier decision to justify politically an economic development in the green belt than housing land release. Concern about the relationship between the NPPF, housing land and the green belt led to new guidance being published in October 2014. This guidance reiterated that green belt boundaries should be altered only in exceptional circumstances, so did not change government policy. But it did stress that housing and economic needs did not automatically outweigh policies of restraint such as green belts (DCLG *et al.*, 2014). It was seen by some as at least in some measure a political move (Tant, 2014), perhaps partly prompted by the May 2014 local elections in England, when the importance of green belts as a political issue was reaffirmed. In these elections, the United Kingdom Independence Party (UKIP) gained 163 council seats; the protection of the green belt had been identified as a key campaign issue across the South East, where many of these seats were won (Geoghegan, 2014b).

In response to the October 2014 guidance, several local authorities announced that they were reviewing their draft local plans that had contained green belt deletions (Carpenter, 2014; Geoghegan, 2014a), so, despite suggestions that the guidance had actually changed very little (Tant, 2014), it evidently had some effect. The Housing and Planning Minister also wrote to the Mayor of London in January 2015, in response to the latter's request to publish alterations to the London Plan. His letter stated that, while he endorsed the publication of the alterations to the plan, the Mayor's letter raised 'a few issues', and it went on to stress that 'the Green Belt should be given the highest protection in the planning system and is an environmental constraint which may impact on the ability of authorities to meet their housing need' (Lewis, 2015).

These strong affirmations of the importance of the green belt may have had some degree of impact on the results of the 2015 general and local elections. UKIP maintained its strong rhetoric on green belts, emphasising in its manifesto that it would replace the NPPF in order to 'genuinely protect the green belt' (UKIP, 2015, p. 35). The Conservative Party manifesto similarly promised to protect the green belt, but no such commitments were made in the Labour or Liberal Democrat manifestos. Single-issue green belt candidates also stood in some parts of the country. Susan Parker, the leader of the Guildford Greenbelt Group, was a candidate for the Guildford constituency and, in the event, received only 538 votes, 1 per cent of the total, while the sitting Conservative Party MP increased her majority by nearly 4 per cent (BBC, 2015). Parker, however, was one of three Guildford Greenbelt Group candidates elected to Guildford Borough Council, and Conservative candidates also did well in other parts of the country in the local (as well as national) elections. This was ascribed partly to the support for green belts in the party's manifesto, leading to concerns from some that development in the green belt might have become less likely in those areas (Marrs, 2015).

The scale of development on green belts has certainly remained very low in recent years. Only 3 per cent of new houses in England were built in the green belt in 2013–14 (Carpenter, 2015), and there were substantial variations between local authorities. Some, with high proportions of green belt land, predictably saw a much higher proportion of housing, an example being Three Rivers District Council in Hertfordshire, where 76 per cent of new houses in 2013–14 were built on green belt land – though green belt makes up 77 per cent of the district, constraining its options somewhat. In addition, that 76 per cent consisted of only 142 houses (Three Rivers District Council, 2014), suggesting the overall impact on the green belt is small. The approach taken by local authorities in the production of local plans has varied, some ruling out large-scale green belt release (Agbonlahor, 2015), others concluding that such releases are necessary. Newcastle City Council, for example, removed another 410 hectares from its green belt in its 2015 plan (Sell, 2015), following up its previous deletion in the early 2000s for the Great Park (see above). The overall amount of green belt land has remained broadly constant, with the figure in 2015 put at 16,366 square kilometres (DCLG, 2015b). This is a slight decrease from the 1997 figure of 16,523 square kilometres, but the government is keen to stress that there has in fact been an increase of 320 square kilometres in the area of green belts during that time, and the loss is due to some green belt being redesignated as National Park, which 'confers a higher status of protection in relation to landscape and scenic beauty than Green Belt' (DCLG, 2015a). This suggests strong political support for green belts at the time of writing, itself no surprise in the context of the previous sixty years, where such support has been more or less constant.

As we noted in the introduction to this chapter, what has been less constant is any view about the purpose, role, objective, aim or otherwise of green belts, since Ebenezer Howard, and those who inspired him, conceived of them primarily as agricultural reserves. In 1962 an American researcher, looking into the applicability of green belts to the American context, noted that 'No two [English] planners gave the same explanation of the underlying rationale behind green-belt policy' (Mandelker, 1962, p. 27). This is perhaps unsurprising, given the wide range of rationales put forward over the last 120 years or more. Table 2.2 summarises some of what we have discussed in this chapter on this specific subject, including the initial ideas of Howard and others, and how official policy has changed in England.

Table 2.2 Changing (explicit or implicit) purposes of green belts in England

Year	Source	Purpose(s) of green belts
1898	Ebenezer Howard	Primarily agriculture
1929	Raymond Unwin	Agriculture and recreation
1920s–1930s	GLRPC	Through archive research, Amati and Yokohari (2006) show the GLRPC approached the National Playing Fields Association in 1924, arguing the green belt could be used for playing fields; the Treasury in 1934; and both the Air and Army Ministry, with the aim of using the green belt for aerodromes and barracks.
1942	Scott report	Primarily agriculture, also recreation
1942 and 1946	Uthwatt and Reith reports	Urban containment and agriculture
1955	MHLG Circular 42/55	Green belts can be established 'in order: (a) to check the further growth of a large built-up area; (b) to prevent neighbouring towns from merging into one another; or (c) to preserve the special character of a town. Wherever practicable, a Green Belt should be several miles wide, so as to ensure an appreciable rural zone all round the built-up area concerned ...' (paras 3–4).
1957	MHLG Circular 50/57	The green belt 'inner boundary ... will mark a long-term boundary for development' (p. 1).
1960	MHLG Circular 37/60	'The aim should be to encourage employment as well as people to move out from the large cities to places beyond the green belt' (p. 2).
1984	DoE Circular 14/84	Green belts 'have a broad and positive planning role in checking the unrestricted sprawl of built-up areas, safeguarding the surrounding countryside from further encroachment, and assisting in urban regeneration ... The Government reaffirms the objectives of Green Belt policy and the related development control policies set out in Ministry of Housing and Local Government Circular 42/55 ...' (p. 1).
1988	PPG2	'Green Belts have five purposes: • to check the unrestricted sprawl of large built-up areas; • to safeguard the surrounding countryside from further encroachment; • to prevent neighbouring towns from merging into one another; • to preserve the special character of historic towns; and • to assist in urban regeneration. Green Belts also have a positive role in providing access to open countryside for the urban population' (paras 4–5).

(Continued)

Table 2.2 (Continued)

Year	Source	Purpose(s) of green belts
1995	PPG2	'The fundamental aim of Green Belt policy is to prevent urban sprawl by keeping land permanently open ... There are five purposes of including land in Green Belts [as above] ... Once Green Belts have been defined, the use of land in them has a positive role to play in fulfilling the following objectives: <ul><li>to provide opportunities for access to the open countryside for the urban population;</li><li>to provide opportunities for outdoor sport and outdoor recreation near urban areas;</li><li>to retain attractive landscapes, and enhance landscapes, near to where people live;</li><li>to improve damaged and derelict land around towns;</li><li>to secure nature conservation interest; and</li><li>to retain land in agricultural, forestry and related uses' (paras 1.4–1.7).</li></ul>
2012	NPPF	'The fundamental aim of Green Belt policy is to prevent urban sprawl by keeping land permanently open; the essential characteristics of Green Belts are their openness and their permanence. Green Belt serves five purposes: <ul><li>to check the unrestricted sprawl of large built-up areas;</li><li>to prevent neighbouring towns merging into one another;</li><li>to assist in safeguarding the countryside from encroachment;</li><li>to preserve the setting and special character of historic towns; and</li><li>to assist in urban regeneration, by encouraging the recycling of derelict and other urban land' (p. 19).</li></ul>

Other nations in the UK

This chapter has focused on the evolution of green belt policy and practice in England, in part because of the substantially larger amount of literature available on that nation, which itself is perhaps because 'Generally a less absolutist Green Belt is sought in the Celtic countries' (Elson and Macdonald, 1997, p. 176). A less "absolutist", or rigid, policy is one that is less likely to be controversial and has more scope for adaptation and pragmatism in practice, which certainly seems to be the case in Scotland at least (Ward, 2004). Elson and Macdonald argued that the *use* of green belt land is also more of a focus in the Celtic countries – guidance in England, particularly in more recent years, has stressed that, regardless of the use to which it is put, green belt is aimed primarily at limiting urban sprawl. The landscape, recreation and economic possibilities of green belt are, in contrast, important factors in Scotland, Wales and Northern Ireland – perhaps, Elson and Macdonald speculate, because more of the countryside is accessible in those countries than in the South of England, so the public is less concerned with protecting land around cities that may not be attractive or in positive use. In the following subsections we briefly review how green belt policy has developed in Scotland, Wales and Northern Ireland.

Scotland

Lloyd and Peel (2007, p. 643) believe that evidence of something akin to a green belt around settlements in Scotland, for agricultural, military, recreation or disease prevention purposes, 'can be traced to ancient times'. As with England, however, it was the advisory plans of Patrick Abercrombie that played a significant role in developing green belt policy in the middle of the twentieth century. Abercrombie published advisory plans for the Clyde Valley, including Glasgow, in 1946 and for 'The City and Royal Burgh of Edinburgh' in 1949. He proposed green belts for both, following the recommendation of advisory committees set up during the Second World War. The Glasgow plan, perhaps because it covered a more extensive area, had a wider, more ambitious green belt, which Bramley *et al.* (2004) observe allowed it to shape the broader city-region – unlike the Edinburgh example, which is narrower and stops short of the settlements beyond the city, which even then were functionally part of its city-region. The view of Bramley and his colleagues was that this limits the potential of the Edinburgh green belt to work at the city-region scale. Mandelker (1962, p. 153) argues that the Edinburgh green belt was deliberately designed to be narrow because 'The Scottish department [of the UK government] does not feel that the holding of a permanent green belt stretching over wide areas is feasible' – instead using the green belt to demarcate an area of open space close to the city centre. Such an attitude on the part of 'the Scottish department' might also explain only the partial incorporation of the Clyde Valley green belt into development plans (Ward, 2004), unlike the Edinburgh example, which was incorporated into the 1953 Edinburgh Development Plan, a cooperative effort on the part of local planning authorities (Skinner, 1976). The objectives of the latter were to limit the expansion of the city, to prevent coalescence, to preserve agricultural land, and to preserve and enhance the setting of the city – similar to those identified in the English context, but apparently derived separately (ibid.).

In 1960 the use of green belts and green wedges was formalised in Scotland through the Department of Health for Scotland (DHS) Circular 40/60, which drew on MHLG circulars 42/55 and 50/57 but did not take forward the idea of "white land" from the latter (Lloyd and Peel, 2007). DHS Circular 40/60 recognised 'the basic problem inherent in Green Belt policies' (Skinner, 1976, p. 10) – that of development "leapfrogging" over green belts to land beyond – so also introduced a general principle of restricting housebuilding in the countryside. The green belt was intended to be primarily a rural area, where 'the normal life and pursuits of the countryside may be maintained' (cited in Lloyd and Peel, 2007, p. 652), and not purely negative – rather, the 'amenity' of the land should be improved. In 1962 the Scottish Development Department (SDD) issued Circular 2/62, which encouraged the development of Areas of Great Landscape Value. These were not green belts, but Skinner (1976) notes they have often functioned in a similar way and have been promoted around smaller settlements such as Perth and Dumfries.

There was a varying degree of enthusiasm for green belts on the part of local authorities in Scotland: three of the six original Scottish green belts were not included in the draft development plans produced from 1953 onwards. The additional ones were included either at the request of the Secretary of State or because constituent local authorities requested it after submission (Skinner, 1976). This variable attitude was also reflected, Skinner notes, in how the green belts were used in development control decisions subsequent to their establishment. At the time he wrote his study, green belts in Scotland were 333,000 acres in size (134,532 hectares), 2 per cent of the land in Scotland. The Minister of the Environment, in 1974, proposed releasing small areas of green belt which were no longer functioning well in order to 'pursue even more vigorously the policy of preserving and enhancing effective areas of Green Belt' (ibid., p. 32).

This more sceptical attitude to green belts was continued into the 1985 SDD Circular 24/1985, which, although it 'carries over three "main purposes"' (Bramley *et al.*, 2004, p. 20) of green belts from Circular 40/60 (see below), called for realism and a reflection upon whether green belt releases were desirable in order to maintain depopulating rural settlements and accommodate the need for development (Elson and Macdonald, 1997; Lloyd and Peel, 2007).

The three 'main purposes' of Scottish green belts were specified as being:

i   to maintain the identity of towns by establishing a clear definition of their physical boundaries and preventing coalescence;
ii   to provide countryside for recreation or institutional purposes of various kinds; or
iii to maintain the landscape setting of towns.

(Scottish Office, 1985, Annex)

These objectives have been adopted to varying degrees by local authorities over the duration of Scottish green belt policy (Skinner, 1976), with others, including assisting urban regeneration, being cited by some local authorities (Bramley *et al.*, 2004), Glasgow among them (Elson and Macdonald, 1997). The Glasgow green belt has also seen substantial deletions and extensions in iterations of its structure plan but remains by far the largest. Dundee abandoned its formal green belt in 1978, while Stirling introduced one in 1999 and St Andrews in 2012.

In 2006 *Scottish Planning Policy 21: Green Belts* formalised the urban regeneration objective of green belt policy in Scotland, specifying the following objectives:

i   to direct planned growth to the most appropriate locations and support regeneration;
ii   to protect and enhance the character, landscape setting and identity of towns and cities; and
iii to protect and give access to open space within and around towns and cities, as part of the wider structure of green space.

(Scottish Executive Development Department, 2006, para. 6)

Table 2.3 Changing (explicit or implicit) purposes of green belts in Scotland

Year	Source	Purpose(s) of green belts
1960	DHS 40/1960	'experience suggests that urban sprawl can best be controlled by the formal designation in Development Plans of clearly defined green belts around and between existing built up areas ... a green belt should be regarded, and should be preserved, essentially as a rural area in which the normal life and pursuits of the countryside may be maintained' (cited in Lloyd and Peel 2007, p. 652).
1985	SDD 24/1985	'Three main purposes: i to maintain the identity of towns by establishing a clear definition of their physical boundaries and preventing coalescence; ii to provide countryside for recreation or institutional purposes of various kinds; or iii to maintain the landscape setting of towns'.
1996	NPPG11 Sport, Physical Recreation and Open Space	'states that "Arising from these [above] purposes a number of ancillary purposes have become commonly accepted, namely to reduce the need to travel, and to assist in urban regeneration through the containment of urban areas and directing development to inner city areas" (para. 57)' (Bramley *et al.* 2004, p. 20).
2006, 2010, 2014	SPP21 and two iterations of Scottish Planning Policy	'the key objectives of green belt policy are: i to direct planned growth to the most appropriate locations and support regeneration; ii to protect and enhance the character, landscape setting and identity of towns and cities; and iii to protect and give access to open space within and around towns and cities, as part of the wider structure of green space' (para 6).

The objective to prevent coalescence was dropped from the 1985 list at this point. The guidance does state that green belts can be used to prevent coalescence but is again pragmatic in noting that, in some cases, coalescence could be the most sustainable development pattern. Similarly, it states that in many cases a green belt is not necessary. This policy has remained in place in the 2010 and 2014 combined Scottish Planning Policy (Scottish Government, 2010, 2014). Table 2.3 summarises the changes to Scottish policy on green belts.

Wales

The history of formal green belts in Wales is much more recent than in England and Scotland. Though proposals were put forward for green belts in Wales in the 1950s (Tewdwr-Jones, 1997), it was only in 1990, with the launch of the UK-wide White Paper *This Common Inheritance*, that the government proposed statutory green belts in Wales. The SoS for Wales asked local authorities to consider them both for the purpose of protecting the countryside and to help accommodate future development (Elson and Macdonald, 1997). Both before and after that intervention, many local authorities rejected the creation of green belts, believing that economic growth was more important than limiting urban sprawl (ibid.). This position was reinforced in both the draft and final Planning Policy Guidance for Wales (in 1995 and 1996 respectively), which were positive towards development and discouraged green belts unless local authorities could demonstrate that other policies for managing development were inadequate (Elson and Macdonald, 1997; Tewdwr-Jones, 1997).

The Newport Unitary Development Plan 1996–2011 (adopted in 2006) contained the first (and still only) formal green belt in Wales, though it 'followed earlier regional collaboration through SEWSPG [South East Wales Strategic Planning Group] on Green Belt issues around the Capital' (Newport City Council, 2015, p. 13). Other local authorities in Wales, such as Gwent and Clwyd, have achieved similar objectives to green belts by using green wedges and green barriers (Elson and Macdonald, 1997; Ward, 2004).

The most recent policy guidance, issued in 2002 and maintained in the latest published consolidated planning policy, sets out that the *purpose* of green belts in Wales is to:

• prevent the coalescence of large towns and cities with other settlements;
• manage urban form through controlled expansion of urban areas;
• assist in safeguarding the countryside from encroachment;
• protect the setting of an urban area; and
• assist in urban regeneration by encouraging the recycling of derelict and other urban land.

(Welsh Government, 2014, p. 54)

It will be noted that these are more positive elements than those in England, and arguably also those in Scotland. As with the 1996 guidance, this iteration stresses that local authorities must demonstrate why 'normal' planning policies are inadequate. Table 2.4 illustrates the evolution of green belt policy in Wales.

Northern Ireland

Northern Ireland 'remained largely a rural society until the latter part of the nineteenth century' (Sterrett and Bassett, 2005, p. 144), and it has been said that the romantic notion of the countryside predominant in England was transferred to Northern Ireland and influenced early attempts to institute planning (McEldowney, 2005). In January 1942, the Royal Society of Ulster Architects argued for the introduction of green belts across the province (Murray, 1991), and in February of the same year a report on Belfast presented to the Minister of Home Affairs by W. R. Davidge, a past president of the (R)TPI, recommended the same. A Planning Commission was set up which, in 1945, developed the proposals further for Belfast and the adjoining parts of the counties of Antrim and Down. The compensation that had to be paid to landowners by local authorities,

Table 2.4 Changing (explicit or implicit) purposes of green belts in Wales

Year	Source	Purpose(s) of green belts
1990	David Hunt MP, SoS for Wales	Should 'not just be about protecting the countryside but defining settlement patterns and urban structure, helping to accommodate future jobs and housing' (Elson and Macdonald 1997, p. 170).
1995	Consultation draft, Planning Policy Guidance (Wales)	'The purposes of Green Belts set out in the draft guidance differ slightly from those for England, particularly in not mentioning sprawl but advocating protecting the setting of all urban areas not just those of historic towns' (Elson and Macdonald 1997, p. 176).
2002–2014	Planning Policy Wales	'The purpose of a Green Belt is to: <ul><li>prevent the coalescence of large towns and cities with other settlements;</li><li>manage urban form through controlled expansion of urban areas;</li><li>assist in safeguarding the countryside from encroachment;</li><li>protect the setting of an urban area; and</li><li>assist in urban regeneration by encouraging the recycling of derelict and other urban land.</li></ul> … When including Green Belt policies in their plans, authorities must demonstrate why normal planning and development management policies would not provide the necessary protection' (pp. 64–5).

and a reluctance to limit economic growth, initially meant that local authorities were not keen to place very strict controls on development. There was a need for 24,600 new houses in and around Belfast, so the Planning Commission deleted 3,000 acres of green belt in its 1951 proposals. Belfast Corporation at this time also wanted to expand its boundaries, but this was prevented. By 1959 Belfast 'remained a sprawling and congested city' (Murray, 1991, p. 100).

In 1960 the architect Sir Robert Matthew was commissioned to prepare a report on planning in the Belfast region. His final report, published in 1963, was based around urban containment and rural conservation (Murray, 1991, 2005; Sterrett and Bassett, 2005), including a "stopline" to limit the growth of the city. The latter did not include a specific width of green belt, and instead merely identified where growth should stop, with 'a presumption against development in the greenscape beyond' (Murray, 1991, p. 105). The principles of urban containment formed the basis of a 1964 ministerial circular regarding planning across the whole of Northern Ireland, despite the fact that no empirical evidence existed to confirm their applicability outside Belfast (Murray, 1991). The urban focus of Matthew's report and the subsequent circular, while similar in tone to the green belt policy in England, were criticised for neglecting the rural, a particular issue in Northern Ireland given its 'unique and complex inheritance of history and culture' (cited in Murray, 2005, p. 171).

This complex inheritance is perhaps exemplified in the topic of dispersed or isolated rural housing, something which has been identified as being particularly popular in Northern Ireland (Murray, 2010). While green belt policy and implementation was more or less consistent from the 1960s onwards (Murray, 1991), policy on dispersed housing has shifted over those years (Murray, 2005), often in response to resentment at policies such as the 1975–95 *Regional Physical Development Strategy*, which proposed rural population decline and urban population growth.

DPO Circular 37/85 defined a wide range of objectives of green belt policy in Northern Ireland:

1   to prevent urban coalescence;
2   to prevent ribbon development and urban sprawl;
3   to contribute to urban regeneration;
4   to facilitate best use of existing infrastructure and minimise cost of new infrastructure provision;
5   to create a long term land bank for future town expansion;
6   to protect areas of high amenity and recreation value within or adjacent to urban areas;
7   to protect good farmland taking into account boundaries of agricultural holdings;
8   to protect natural resources such as minerals and water catchment areas;
9   to create a clear division between town and country;
10   to preserve the special character of a town.

(Cited in Murray, 1991, pp. 165–6)

By 1991, there was 'virtually complete coverage by statutory development plans', with accompanying green belts, in Northern Ireland (ibid., p. 77). Despite several of the objectives above relating to the quality of green belt land, Murray notes that the Department of Environment was primarily concerned with green belt as a "stopper". The *Planning Strategy for Rural Northern Ireland* (1993) was seen as a way to discourage rural housing, 'not least in relation to the designation of extensive Countryside Policy Areas and Green Belts' (Murray, 2005, p. 176). It focused the 'strategic objectives' of green belts somewhat, from the ten listed in the 1985 guidance to five:

• to prevent the unrestricted sprawl of large built-up areas;
• to prevent neighbouring settlements from merging;
• to safeguard the surrounding countryside;
• to protect the setting of settlements; and
• to assist in urban regeneration.

(DoE (Northern Ireland), 1993)

However, as at 2005, 'The number of houses being built in the northern Irish countryside continues to increase' (Sterrett and Bassett, 2005, p. 149), so 'More recent times have witnessed a serious attempt by Planning Service and the Department

Table 2.5 Changing (explicit or implicit) purposes of green belts in Northern Ireland

Year	Source	Purpose(s) of green belts
1963 and 1964	Matthew report and ministerial circular	Urban containment
1985	DPO Circular 37/85: Areas of Outstanding Natural Beauty and Green Belts	'The objectives of green belt are: 1 to prevent urban coalescence; 2 to prevent ribbon development and urban sprawl; 3 to contribute to urban regeneration; 4 to facilitate best use of existing infrastructure and minimise cost of new infrastructure provision; 5 to create a long term land bank for future town expansion; 6 to protect areas of high amenity and recreation value within or adjacent to urban areas; 7 to protect good farmland taking into account boundaries of agricultural holdings; 8 to protect natural resources such as minerals and water catchment areas; 9 to create a clear division between town and country; 10 to preserve the special character of a town' (Murray 1991, pp. 165–6).
1993	*Planning Strategy for Rural Northern Ireland*	'Where it is considered necessary to protect landscapes from excessive or inappropriate development, Green Belts will be designated around cities and towns.' 'The "strategic objectives" … • to prevent the unrestricted sprawl of large built-up areas; • to prevent neighbouring settlements from merging; • to safeguard the surrounding countryside; • to protect the setting of settlements; and • to assist in urban regeneration. In addition to its strategic role in restraining development pressures it is important that the open countryside in the Green Belt makes a positive contribution to meeting the outdoor recreation needs of the urban population. Green Belt designation also provides the opportunity for the enhancement of urban fringe areas in the knowledge there is a long-term commitment to retain their open character.'

for regional Development to claw back the presumption in favour of development by expanding green belt coverage and introducing new spatially extensive countryside Policy areas in area Plans' (Murray, 2010, pp. 45–6). Table 2.5 shows how green belt policy in Northern Ireland has developed.

Conclusion

One thing the tables concluding the preceding four subsections show is that attempting to clarify the intended purpose of green belts now, at the end of the nineteenth century, or at any point in between, is not straightforward. In Chapter 1 we quoted Professor Sir Peter Hall and his colleagues, who identified what they saw as a series of negative and regressive consequences of the operation of the English planning system in the years following its inception in 1947. They concluded that 'None of this was in the minds of the founding fathers of the planning system' (Hall *et al.*, 1973b, p. 433). As planners ourselves, we would like to believe that is true, and it is certainly the case that Ebenezer Howard, for example, was progressive in outlook and genuinely sought to improve the lot of what he would have called "the working man". But it is equally the case that Howard's conception had been almost entirely lost by the time that green belts became part of official government policy in England in 1955. The intentions of other "founding fathers" of planning, such as Patrick Abercrombie, is not entirely clear – we have noted above Abercrombie's endorsement, through his position in the CPRE, of the romantic, aesthetic-focused critiques of inter-war urbanisation espoused by Clough Williams-Ellis and others. That is not to imply that Abercrombie had regressive intentions in his contributions to the Barlow Commission report in 1940 or in his famous advisory plans for London, Glasgow, etc. If any fault can be laid at the feet of Abercrombie, it is, firstly, a failure, along with everyone else, to anticipate the massive changes that would take place in London and the UK in the twenty years after the Second World War – the growth in population and car ownership perhaps the most profound in this context – and, secondly, to neglect a consideration of how his proposals for London and elsewhere would be implemented in a context of diminished political will for national or regional planning and state-led development.

As we will discuss, specifically in Chapter 3, the impacts of green belts in this dramatically different context from that envisaged by Abercrombie have been profound and, as argued by Peter Hall and others, regressive in nature. The costs of the green belt policy were not foreseen at the time of implementation, and the policy has not been fundamentally changed since its inception. According to Evans, 'What would appear to have happened is that what was thought of at the time as a prediction [that birth rates would stay low, and that (manufacturing) jobs would be decentralised] became, as circumstances changed, a control' (1985, p. 203). Before we move on, it is worth reflecting on why this might be the case and why green belts have proved so enduringly popular with some of the stakeholders we identified in Chapter 1.

Firstly is the fact that green belts can, in some sense, be seen as "all things to all men", partly because there have been many and various views (publicly stated or not) on the part of governments as to what their purpose is. Whether the aim/ purpose of green belts is to limit urban sprawl, provide access to the countryside for urban dwellers, protect agricultural land, protect and enhance biodiversity, or something else, many individuals and bodies, public and private, can find something to support. This adaptability has been identified as the key reason for the continuing success of green belts (Murray, 1991), typical of the planning system in being adaptive to change (Elson, 1986), and, perhaps critically, typical of change in Britain in general, which 'is often handled artfully – as if it were not really change. Continuity is maintained by periodic reaffirmation of policy and by explicit assertions that the situation is covered by existing policy' (Foley, 1963, p. 160). That statement is as true today as it was in 1963, and perhaps suggests that any change to green belt policy in the future needs to be considered in such terms.

Secondly, planners themselves like the green belt. It is simple and well understood by the public. Some (Elson, 1986) have argued that alternative policies that might allow a more nuanced and pragmatic approach would by definition be more complex and harder to explain, justify and defend. It has also been argued that planners view the success of the green belt as being in some way tied to the success of the profession on account of their close historical intertwining. The President of the Town Planning Institute in 1955 argued that the green belts were the '"very *raison d'être*"' of the planning system (cited in Amati, 2008, p. 6), and similar views would not be hard to find today. Evans saw the reluctance to amend the green belt policy as being partly down to 'a conflict between the relativism of the economist and the absolutism of the planner' (1985, p. 203). He stated that an economist, looking at costs and benefits, would have argued for flexibility and some green belt release, preserving recreational land but bringing costs in line with benefits.

> Planners, on the other hand, having laid down the policy appear to have regarded it as fixed, to regard the green belts as first defined to be the best attainable, and to think that, since it is the best, there is little point in thinking of modifying it.
>
> (Ibid.)

Thirdly, green belts are popular with politicians. This applies both to those at the national level, largely for electoral reasons (see below), and to those at the local level, for the same reasons and because the very strong presumption against development inherent in green belt designation gives local authorities a great deal of power in their dealings with developers and central government (Elson, 1986).

Finally, and most importantly, the 'general public', if they can be lumped together in that way, consistently show support for the concept of green belts. Mandelker (1962) thought this was due to a number of factors, including the historic romanticism with which the countryside is viewed by the English[7] and the twin symbolism of the Green Man and circles. The former 'was a spring sacrifice, poor fellow. They'd wrap him in leaves and boughs and drown him in the river', while the latter was 'a ritual symbol throughout the ages. Have you seen Stonehenge?' ('an English planning officer', cited ibid., p. 2). Whether this a romantic step too far we are not sure, but there is a broad acceptance that the English do equate the countryside with the romantic (see Matless, 1998, among others), so protecting it from development becomes part of national identity. This is reflected in how the language used to describe green belt classifies it 'as *good*. There are encroachments upon it, inroads into it, incursions into it. It is nibbled into, sacrificed, desecrated … the belt is inviolate, sacred, sacrosanct, protected and defended' (Warren-Evans, 1974, p. 20). This popularity with the general public is reflected in, and reflects, the same status with which green belt is viewed by national politicians, who, as the examples of Conservative Secretaries of State from the 1980s show, risk development on the green belt at their peril.

There may be signs, however, that public support for green belts has fallen in the last ten years. In August 2015, to coincide with the sixtieth anniversary of Circular 42/55, the CPRE published the results of a survey it had commissioned from the polling organisation Ipsos MORI. That survey asked 'To what extent do you agree or disagree, in principle, that existing Green Belt land in England should be retained and not built on?', with which 64 per cent of respondents strongly agreed or tended to agree (Ipsos MORI, 2015). However, in a very similar survey ten years earlier, which asked the question 'How much, if at all, do you agree or disagree that green belt land should remain open and undeveloped, and building on it not allowed?', 84 per cent strongly agreed or tended to agree (Ipsos MORI, 2005). That suggests a 20 per cent fall in support for the green belt in ten years, not something that features either in the CPRE's press release accompanying the report (CPRE, 2015) or in press reports at the time (Press Association, 2015). An article in *The Guardian* (Golby, 2015), reacting to the 2015 poll, pointed out the lower level of support for green belts among younger people, and argued that this was because of their inability to buy a house in cities such as London, which in turn many blame on the constraints of the planning system (see Chapter 3 for more on these arguments). If there is such a generational shift in attitudes, then the longer-term prospects for green belt reform may be more likely than they have been since its inception.

Notes

1 Translated by Osborn (1946, p. 170) as 'Designated Field … a belt of open space'.
2 The GLRPC was set up by Neville Chamberlain, then Minister of Health, in 1927: 'he asked them, among other things … to consider whether London should "be provided with something which might be called an agricultural belt, as has often been suggested"' (MHLG, 1962, p. 2).
3 W.C. = water closet = lavatory.
4 F. J. Osborn being Chairman of the TCPA – this again illustrates the proximity and relationship between some of the key actors in the story of the green belt.
5 Crossman here perhaps references the famous Copenhagen 'finger plan' (Vejre *et al.*, 2007).
6 Eventually Cheshire County Council was abolished and replaced in 2009 by Cheshire West & Chester and Cheshire East councils.
7 And to a lesser extent the other parts of the UK.

References

Abercrombie, P. (1945) *Greater London Plan* (London: HMSO).

Agbonlahor, W. (2015) Oxfordshire council rules out large-scale green belt release, 14 October, www.planningresource.co.uk/article/1368415/oxfordshire-council-rules-large-scale-green-belt-release.

Amati, M. (2008) Green belts: a twentieth-century planning experiment, in M. Amati (ed.), *Urban Green Belts in the Twenty-first Century* (Aldershot: Ashgate), pp. 1–17.

Amati, M., and Yokohari, M. (2006) Temporal changes and local variations in the functions of London's green belt, *Landscape and Urban Planning*, 75(1–2): 125–42.

Amati, M., and Yokohari, M. (2007) The establishment of the London greenbelt: reaching consensus over purchasing land, *Journal of Planning History*, 6(4): 311–37.

Anonymous (1956) Ye olde English green belt, *Journal of the Town Planning Institute*, 42: 68–9.

Bairoch, P., and Goertz, G. (1986) Factors of urbanisation in the nineteenth century developed countries: a descriptive and econometric analysis, *Urban Studies*, 23: 285–305.

Barker, K. (2004) *Delivering Stability: Securing our Future Housing Needs* (London: HMSO).

Barlow, Sir M. (1940) *Report [of the] Royal Commission on the Distribution of the Industrial Population* (London: HMSO).

BBC (2015) Guildford parliamentary constituency results, www.bbc.co.uk/news/politics/constituencies/E14000719.

Beach Thomas, W. (1938) The Home Counties, in C. Williams-Ellis (ed.), *Britain and the Beast* (Letchworth: Temple Press), pp. 200–11.

Bramley, G., Hague, C., Kirk, K., Prior, A., Raemaekers, J., Smith, H., Robinson, A., and Bushnell, R. (2004) *Review of Green Belt Policy in Scotland* (Edinburgh: Scottish Executive Social Research).

Carpenter, J. (2014) Green belt guidance update prompts council to reassess local plan, 26 November, www.planningresource.co.uk/article/1323914/green-belt-guidance-update-prompts-council-reassess-local-plan.

Carpenter, J. (2015) Data blog: The councils that saw the most green belt housing growth in 2013/14, 7 August, www.planningresource.co.uk/article/1359195/data-blog-councils-saw-green-belt-housing-growth-2013-14.

Clapp, J. A. (1971) *New Towns and Urban Policy – Planning Metropolitan Growth* (New York and London: Dunellen).

Cobbett, W. (1885) *Rural Rides*, Vol. 1 (London: Reeves & Turner).

Conservative Party (2010) *Invitation to Join the Government of Britain: The Conservative Manifesto 2010* (London: Conservative Party).

CPRE (2013) *Green Belt and the National Planning Policy Framework: 18 Months On*, www.cpre.org.uk/resources/housing-and-planning/green-belts/item/3400-green-belts-and-the-national-planning-policy-framework-18-months-on.

CPRE (2015) 60th anniversary poll shows clear support for green belt, www.cpre.org.uk/media-centre/latest-news-releases/item/4033-60th-anniversary-poll-shows-clear-support-for-green-belt.

Cullingworth, B. (1980) *Land Values, Compensation and Betterment*, Vol. IV (London: HMSO).

Cullingworth, B., and Nadin, V. (2006) *Town and Country Planning in the UK* (14th ed., London and New York: Routledge).

Cullingworth, B., Nadin, V., Hart, T., Davoudi, S., Pendlebury, J., Vigar, G., Webb, D., and Townshend, T. (2015) *Town and Country Planning in the UK* (15th ed., London: Routledge).

DCLG (2012) *National Planning Policy Framework* (London: Department for Communities and Local Government).

DCLG (2015a) *Local Authority Green Belt Statistics: Annex 1*, https://www.gov.uk/government/uploads/system/uploads/attachment_data/file/464783/Annex_1_-_Green_Belt_Statistics_2014-15_Tables.xlsx.

DCLG (2015b) *Local Planning Authority Green Belt: England 2014/15* (London: Department for Communities and Local Government).

DCLG, Lewis, B., and Pickles, E. (2014) Councils must protect our precious green belt land, Press release, https://www.gov.uk/government/news/councils-must-protect-our-precious-green-belt-land.

DETR (2000) *Planning Policy Guidance Note 3: Housing* ([rev. ed.,] London; The Stationery Office).

DoE (1973) *Land Availability for Housing*, Circular 122/73 (London: HMSO).

DoE (1984) *Green Belts*: Circular 14/84 (London: HMSO).

DoE (1988a) *The Green Belts* (London: HMSO).

DoE (1988b) *Planning Policy Guidance 2: Green Belts* (London: HMSO).

DoE (1993) *The Effectiveness of Green Belts* (London: HMSO).

DoE (1995) *Planning Policy Guidance 2 (revised): Green Belts* (London: HMSO).

DoE (Northern Ireland) (1993) *Planning Strategy for Rural Northern Ireland: Regional Planning Policies: Policy GB/CPA 1, Designation of Green Belts and Countryside Policy Areas*, www.planningni.gov.uk/index/policy/rural_strategy/psrni_regional_policies/psrni_cpa/psrni_cpa01.htm.

Dunton, J. (2014) Economic case trumps green belt harm, *Planning*, 4 July 2014, pp. 8–9.

Elson, M. J. (1986) *Green Belts: Conflict Mediation in the Urban Fringe* (London: Heinemann).

Elson, M. J., and Macdonald, R. (1997) Urban growth management: distinctive solutions in the Celtic countries, in R. Macdonald and H. Thomas (eds), *Nationality and Planning in Scotland and Wales* (Cardiff: University of Wales Press), pp. 159–80.

Essex, S., and Brayshay, M. (2005) Town versus country in the 1940s: planning the contested space of a city region in the aftermath of the Second World War, *Town Planning Review*, 76(3): 239–64.

Evans, A. (1985) *Urban Economics: An Introduction* (Oxford: Blackwell).

Evelyn, J. ([1661] 2011) *Fumifugium MMXI: A 21st Century Translation of a 17th Century Essay on Air Pollution in London, Sent to King Charles II by the Writer John Evelyn* (London: Environmental Protection UK).

Foley, D. L. (1963) *Controlling London's Growth: Planning the Great Wen 1940–1960* (Berkeley: University of California Press).

Frey, H. (1999) Not green belts but green wedges: the precarious relationship between city and country, *Urban Design International*, 5(1): 13–25.

Geoghegan, J. (2014a) Guidance update puts plan at risk, 17 October, www.planningresource.co.uk/article/1317552/guidance-update-puts-plan-risk.

Geoghegan, J. (2014b) UKIP victors to fight green belt homes, 6 June, www.planningresource.co.uk/article/1297332/ukip-victors-fight-green-belt-homes.

Golby, J. (2015) Why Generation Rent doesn't care about your precious green belt, *The Guardian*, 5 August, www.theguardian.com/commentisfree/2015/aug/05/generation-rent-concrete-over-green-belt-renting?CMP=share_btn_link.

Gunn, S. C. (2007) Green belts: a review of the regions' responses to a changing housing agenda, *Journal of Environmental Planning and Management*, 50(5): 595–616.

Hall, P. (1974) The green belt: its past and future justification, in College of Estate Management (ed.), *The Future of the Green Belt*, Occasional Papers in Estate Management, no. 5 (Reading: College of Estate Management), pp. 1–8.

Hall, P., and Ward, C. (1998) *Sociable Cities: Legacy of Ebenezer Howard* (Chichester: John Wiley).

Hall, P., Gracey, H., Drewitt, R., and Thomas, R. (1973a) *The Containment of Urban England*, Vol. 1 (London: Allen & Unwin).

Hall, P., Gracey, H., Drewitt, R., and Thomas, R. (1973b) *The Containment of Urban England*, Vol. 2 (London: Allen & Unwin).

Hall, P., Hardy, D., and Ward, C. (2003) *To-Morrow: A Peaceful Path to Real Reform: Original Edition with Commentary* (London: Routledge).

HM Government (2010) *The Coalition: Our Programme for Government* (London: HM Government Cabinet Office).

HM Treasury and Office for the Deputy Prime Minister (2005) *The Government's Response to Kate Barker's Review of Housing Supply* (Norwich: HMSO).

Holmans, A. E. (1987) *Housing Policy in Britain* (Beckenham: Croom Helm).

Howard, E. (1898) *To-Morrow: A Peaceful Path to Real Reform* (London: Swann Sonnenschein).

Howard, E. ([1902] 1951) *Garden Cities of To-Morrow*, ed. F. J. Osborn, intro. L. Mumford (London: Faber & Faber).

Humber, R. (1982) This green & pleasant land, *House Builder and Estate Developer: Journal of the Federation of Registered House-Builders*, 41(6): 2–3.

Ipsos MORI (2005) *50th Anniversary of Green Belts*, https://www.ipsos-mori.com/ researchpublications/researcharchive/435/50th-Anniversary-of-Green-Belts.aspx.

Ipsos MORI (2015) *Attitudes towards Green Belt Land: A Study for the Campaign to Protect Rural England*, https://www.ipsos-mori.com/researchpublications/ researcharchive/3611/Attitudes-towards-Green-Belt-land.aspx.

Jackson, J. N. (1972) *The Urban Future* (London: Allen & Unwin).

Joad, C. E. M. (1938) The people's claim, in C. Williams-Ellis (ed.), *Britain and the Beast* (Letchworth: Temple Press), pp. 64–85.

Kaye-Smith, S. (1938) Laughter in the South-East, in C. Williams-Ellis (ed.), *Britain and the Beast* (Letchworth: Temple Press), pp. 32–43.

Labour Party (2010) *A Future Fair for All: The Labour Party Manifesto 2010* (London: Labour Party).

Laws, P. (1955) Ribbon development: present trends, *Town and Country Planning*, 23: 263–8.

Lewis, B. (2015) *Further Alterations to the London Plan*, https://www.london.gov.uk/ what-we-do/planning/london-plan/past-versions-and-alterations-london-plan/further-alterations-london#Stub-111348.

LGiU (2013) Government planning policy puts England's green belts at risk, suggests new research, Press release, Local Government Information Unit, www.lgiu.org.uk/news/ government-planning-policy-puts-englands-green-belts-at-risk-suggests-new-research/.

Liberal Democrat Party (2010) *Change that Works for You, Delivering a Fairer Britain: The Liberal Democrat Manifesto* (London: Liberal Democrat Party).

Lloyd, M. G., and Peel, D. (2007) Green belts in Scotland: towards the modernisation of a traditional concept?, *Journal of Environmental Planning and Management*, 50(5): 639–56.

Lockley, R. M. (1938) The sea coast, in C. Williams-Ellis (ed.), *Britain and the Beast* (Letchworth: Temple Press), pp. 225–39.

Loftus Hare, W. (1937) The green belt – its relation to London's growth, *Journal of the Royal Institute of British Architects*, 44: 677–85.

Lowe, P., and Murdoch, J. (2003) Mediating the 'national' and the 'local' in the environmental policy process: a case study of the CPRE, *Environment and Planning C: Government and Policy*, 21(5): 761–78.

McEldowney, M. (2005) The rural–urban interface: outskirts of European cities, in M. McEldowney, M. Murray, B. Murtagh and J. Sterrett (eds), *Planning in Ireland and Beyond: Multidisciplinary Essays in Honour of John V. Greer* (Belfast: School of Environmental Planning, Queen's University), pp. 319–32.

Mandelker, D. (1962) *Green Belts and Urban Growth: English Town and Country Planning in Action* (Madison: University of Wisconsin Press).

Marrs, C. (2015) Why the local election results could mark a further stumbling block for green belt homes, 15 May, http://www.planningresource.co.uk/article/1347230/why-local-election-results-mark-further-stumbling-block-green-belt-homes.

Matless, D. (1998) *Landscape and Englishness* (London: Reaktion Books).

MHLG (1955) *Green Belts*, Circular 42/55 (London: HMSO).

MHLG (1956) *Report of the Ministry of Housing and Local Government for the Year 1955* (London: HMSO).

MHLG (1957a) *Green Belts*, Circular 50/57 (London: HMSO).

MHLG (1957b) *Report of the Ministry of Housing and Local Government for the Year 1956* (London: HMSO).

MHLG (1960) *The Review of Development Plans*, Circular 37/60 (London: HMSO).

MHLG (1962) *The Green Belts* (London: HMSO).

Miller, M. (1989) The elusive green background: Raymond Unwin and the Greater London regional plan, *Planning Perspectives*, 4(1): 15–44.

More, T. ([1556] 1999) *Utopia* (Boston: Bedford/St Martin's).

Murdoch, J., and Lowe, P. (2003) The preservationist paradox: modernism, environmentalism and the politics of spatial division, *Transactions of the Institute of British Geographers*, 28(3): 318–32.

Murray, M. (1991) *The Politics and Pragmatism of Urban Containment* (Aldershot: Avebury).

Murray, M. (2005) Consultation: new countryside housing and rural planning policy in Northern Ireland, in M. McEldowney, M. Murray, B. Murtagh and J. Sterrett (eds), *Planning in Ireland and Beyond: Multidisciplinary Essays in Honour of John V. Greer* (Belfast: School of Environmental Planning, Queen's University), pp. 169–86.

Murray, M. (2010) *Participatory Rural Planning: Exploring Evidence from Ireland* (Burlington, VT: Ashgate).

Nairn, I. (1955) *Outrage ... a Reprint of the June, 1955, Special Number of the Architectural Review* (London: Architectural Press).

Newport City Council (2015) *Newport Local Development Plan 2011–2026* (Newport: Newport City Council).

Osborn, F. J. (1946) *Green-Belt Cities – the British Contribution* (London: Faber & Faber).

Parsons, K. C. (1990) Clarence Stein and the greenbelt towns: settling for less, *American Planning Association Journal*, 56(2): 161–83.

Pennington, M. (2000) *Planning and the Political Market: Public Choice and the Politics of Government Failure* (London: Athlone Press).

Pepler, G. L. (1911) Greater London, *Transactions of the Royal Institute of British Architects Town Planning Conference*, London, 10–15 October 1910.

Pickles, E. (2012) Written ministerial statement to the House of Commons, 6 September, www.publications.parliament.uk/pa/cm201213/cmhansrd/cm120906/wmstext/120906m0001.htm.

Press Association (2015) Two-thirds of public against building on Britain's green belt land, *The Guardian*, 3 August, www.theguardian.com/politics/2015/aug/03/uk-green-belt-land-survey-countryside-housing.

Reith, J. C. W. (1946) *Final Report of the New Towns Committee* (London: HMSO).

Ridley, N. (1991) *'My Style of Government': The Thatcher Years* (London: Hutchinson).

Ruskin, J. ([1868] 2008) *Sesame and Lilies* (Rockville, MA: Arc Manor).

Sandys, D. (1956) Green belts: minister's reply, *Town and Country Planning*, 24: 151–3.

Scott, L. F. (1942) *Report of the Committee on Land Utilisation in Rural Areas* (London: HMSO).

Scott-Moncrieff, G. (1938) The Scottish scene, in C. Williams-Ellis (ed.), *Britain and the Beast* (Letchworth: Temple Press), pp. 266–78.

Scottish Executive Development Department (2006) *Scottish Planning Policy 21: Green Belts* (Edinburgh: Scottish Executive).

Scottish Government (2010) *Scottish Planning Policy* (Edinburgh: Scottish Government).

Scottish Government (2014) *Scottish Planning Policy* (Edinburgh: Scottish Government).

Scottish Office (1985) *Scottish Development in the Countryside and Green Belts*, Circular 24/85 (London: Scottish Development Department).

Sell, S. (2015) Local plan watch: why more councils are releasing green belt land in their local plans, 9 October, www.planningresource.co.uk/article/1367451/local-plan-watch-why-councils-releasing-green-belt-land-local-plans.

Sharp, E. (1969) *The Ministry of Housing and Local Government* (London: Allen & Unwin).

Short, J. R. (1982) *Housing in Britain: The Post-War Experience* (London: Methuen).

Simmie, J. M. (1974) *Citizens in Conflict: The Sociology of Town Planning* (London: Hutchinson Educational).

Skinner, D. (1976) *A Situation Report on Green Belts in Scotland* (Perth: Countryside Commission for Scotland).

Sorensen, A. (2002) *The Making of Urban Japan: Cities and Planning from Edo to the Twenty-First Century* (London: Routledge).

Stamp, L. D. (1945) Land classification and agriculture, in P. Abercrombie (ed.), *Greater London Plan* (London: HMSO), pp. 86–96.

Sterrett, K., and Bassett, J. (2005) The social shaping of rural vernacular housing, in M. McEldowney, M. Murray, B. Murtagh and J. Sterrett (eds), *Planning in Ireland and Beyond: Multidisciplinary Essays in Honour of John V. Greer* (Belfast: School of Environmental Planning, Queen's University), pp. 139–56.

Sturzaker, J. (2011) Can community empowerment reduce opposition to housing? Evidence from rural England, *Planning Practice and Research*, 26(5): 555–70.

Sturzaker, J., and Shaw, D. (2015) Localism in practice: lessons from a pioneer neighbourhood plan in England, *Town Planning Review*, 86(5): 587–609.

Tant, I. (2014) Briefing: change of tone in reiteration of green belt practice guidance, 17 October, www.planningresource.co.uk/article/1317381/briefing-change-tone-reiteration-green-belt-practice-guidance.

Tewdwr-Jones, M. (1997) Green belts or green wedges for Wales? A flexible approach to planning in the urban periphery, *Regional Studies*, 31(1): 73–7.

Thatcher, M. (1988) Speech to the Royal Society, 27 September, www.margaretthatcher.org/document/107346.

Thatcher, M. (1989) Speech to United Nations General Assembly (Global Environment), 8 November, www.margaretthatcher.org/document/107817.

Three Rivers District Council (2014) *Annual Monitoring Report 2013/14* (Rickmansworth: Three Rivers District Council).

Town and Country Planning Association (1955) Planning commentary, *Town and Country Planning*, 23: 276–80.

Tromans, S. (1991) Review: this common inheritance: Britain's environmental strategy, *Journal of Environmental Law*, 3(1): 168–72.

UKIP (2015) *Believe in Britain: UKIP Manifesto 2015* (Newton Abbot: UK Independence Party).

Unwin, R. ([1909] 1994) *Town Planning in Practice: An Introduction to the Art of Designing Cities and Suburbs* (New York: Princeton Architectural Press).

Uthwatt, A. A. (1942) *Final Report [of the] Expert Committee on Compensation and Betterment* (London: HMSO).

Vejre, H., Primdahl, J., and Brandt, J. (2007) The Copenhagen finger plan, in B. Pedroli, A. van Doorn, G. de Blust, M. L. Paracchini, D. Wascher and F. Bunce (eds), *Europe's Living Landscapes* (Wageningen: Koninkliijke Nederlandse Natuurhitorische Vereniging).

Ward, S. K. (2004) *Planning and Urban Change* (2nd ed., London: Sage).

Warren-Evans, J. R. (1974) The growth of urban areas, in College of Estate Management (ed.), *The Future of the Green Belt*, Occasional Papers in Estate Management, no. 5 (Reading: College of Estate Management), pp. 19–24.

WCED (World Commission on Environment and Development) (1987) *Our Common Future* (Oxford: Oxford University Press).

Welsh Government (2014) *Planning Policy Wales* (7th ed., Cardiff: Welsh Government).

Williams-Ellis, C. ([1928] 1975) *England and the Octopus* (new ed., Portmeirion: Golden Dragon Books).

3   The impacts of green belts in the UK

Introduction

There are a number of ways in which we might assess the impacts of green belts in the UK. As they have been implemented through policies established by the UK, Welsh, Scottish and Northern Irish governments, we could assess them against the purposes laid down in the respective national planning policies, as set out in Tables 2.2 to 2.5. But, as we have already mentioned, there can be a difference between what such "official" policy says and what others interpret as the purposes/objectives/etc. of green belts. There are also ambiguities in government planning policy – for instance, in the previous iteration of policy relating to green belts in England contained in the now superseded *Planning Policy Guidance 2*, in addition to five *purposes* of green belts, six *objectives* were identified that the land within green belts could help fulfil (see Table 2.2). These objectives are no longer part of planning policy relating to England, but this does not mean they are no longer important – indeed, they may be closer to what the general public perceive as the purposes of green belts – for example 'to provide opportunities for access to the open countryside for the urban population' (DoE, 1995, para. 1.7). We have also to consider how we can measure impact. This chapter draws largely upon existing data sources, supplemented with our own study of the green belt around Merseyside, so we are constrained by the data that others have chosen to collect. For instance, many people have explored the impact of green belts on house prices – not something that features as a purpose/objective of green belts in the UK – but we have found very little on whether they have made a contribution to the PPG2 objective to 'improve damaged and derelict land around towns' (ibid.), though a study by the CPRE and Natural England (2010) is one notable exception. In this chapter, therefore, we have gathered evidence in relation to a number of (perceived) impacts that green belts have had or are having.

Another issue concerning data collection is that it is all but impossible in most cases to be certain about the specific impacts green belts themselves are having, for several reasons. Firstly, green belts are just one part of the operation of the planning system. Green belt is just one of the designations placed on land to offer it protection from change – others, with varying degrees of power and with varying objectives, include National Parks, Areas of Outstanding Natural Beauty

(AONBs), Conservation Areas, Sites of Special Scientific Interest (SSSIs), and National and Local Nature Reserves. The Barker review estimated that 36 per cent of land in England was protected from development through such designations, a figure that rose to 60 per cent in the South East (Barker, 2004). There are also other policies of restraint operated by local planning authorities across the country, whether through their own designation of land (for example, as of High Landscape Value or similar) or through a general policy of not allowing housing development outside settlements of a certain size, perhaps because such development is 'unsustainable' (see Sturzaker, 2010, for a discussion of such approaches). Differentiating the impacts of green belts from these other forms of restraint on, for example, house prices is not straightforward. Where authors have bundled the green belt together with other factors (assessing the impact of 'the planning system' is not uncommon), we make that explicit in our discussion of their findings. Secondly, quite beyond the planning system, there are many factors that affect urban development, from the rate of population growth, to the economy, to political change, etc. Again, differentiating the impact of green belts from some of these multifarious complex factors is difficult. That said, for some forms of analysis there are ways of controlling for other factors, and some authors have done just that. Thirdly, with any form of analysis of this type, where we are assessing the impact of a policy in the real world, there is no "null hypothesis" or "control group". Not every town or city in the UK has a green belt – Leicester and Hull being notable examples of those that do not – but comparing these urban areas with similar towns or cities that have green belts is not possible, given the different topographies, histories, collection of other planning policies, etc. We simply cannot know with any degree of certainty what, say, Liverpool would look like without the Merseyside green belt surrounding it – or what Leicester would look like if it *did* have a green belt. We can make an educated guess, and, as we will discuss below, this has been done for London at least, but it is no more than that. We cannot look at any piece of land within the green belt and say, 'If this was not green belt, then it would *definitely* have been developed', nor can we state with any certainty what it might have been developed *for* – housing, industry, etc. It is possible to look at the aggregate scale and try and model the effect of (full or partial) removal of the green belt, or relaxation to a certain extent, and many of the sources we cite in this chapter have taken that approach.

We have grouped the discussion here into a number of subheadings, each dealing with a specific impact or group of impacts.

Impacts on urban sprawl

Why does urban sprawl matter?

In a speech given by Duncan Sandys, Minister for Housing and Local Government, to the House of Commons, he said: 'We have a clear duty to do all we can to prevent the further unrestricted sprawl of the great cities' (MHLG, 1956, p. 55), and this is maintained today, with England's National Planning Policy Framework (NPPF) stating that 'The fundamental aim of Green Belt policy is to prevent urban sprawl by keeping land permanently open' (DCLG, 2012, p. 19). This therefore seems like an appropriate place to start. There are good reasons for limiting urban sprawl. Power (2001) identified a range of impacts and consequent inner-city decline across environmental, social and economic indicators, including, in rural areas, increased road traffic, pollution, energy consumption and noise and, in urban areas, population loss, poverty, empty buildings, loss of shops and amenities, and fragmented neighbourhoods.

During and immediately after the Second World War, protecting agricultural land from urban sprawl was identified as a priority (Scott, 1942; Reith, 1946), and the green belt proposed in Abercrombie's London plan was designed as much for that purpose as any other (Abercrombie, 1945). Between 1934 and 1939, 278,000 acres (111,200 hectares) of agricultural land was built on (Scott, 1942, p. 28). Most observers agree that, where they have been in place, green belts, and the similar policies adopted in Wales (as discussed in Chapter 2), have succeeded in stopping, or at least seriously limiting, physical urban sprawl around the towns and cities of the UK, and hence dramatically slowing the loss of agricultural land that caused such concern to Scott, Abercrombie and others. Detailed quantitative analysis carried out by the geographer Robin Best (Best, 1981; Best and Anderson, 1984), in response to concerns from, among others, Coleman (1977), illustrated this, as shown in Table 3.1. According to his figures, the proportion of agricultural land in Great Britain had fallen only slightly since the Second World War, with the transfer of land to forestry uses having a similar effect as the increase in urban land.

Others have estimated that

the average annual transfer of land from agriculture to urban use in England and Wales [fell from] 25,100 hectares per annum in the 1930s … [to] 17,500 hectares per annum in 1945–50 … 15,000 per annum in 1950–65 … 16,800 in 1965–70 … 9,300 in 1975–80 … [and] 6,500 per annum in 1985–89.

(Ward, 2004, p. 277)

Table 3.1 Percentage of agricultural land in Britain

Great Britain	Population	Agriculture	Woodland	Urban*	Other
1931	44.8 million	79.6	5.6	5.0	9.8
1961	51.3 million	82.0	7.4	7.4	3.2
1971	54.0 million	78.0	8.3	8.2	5.5
1981	54.3 million	76.0	9.4	8.8	5.8

Source: Best (1981, pp. 46–7); Best and Anderson (1984, p. 22).

Note: *Includes all transport land, residential, urban open space, industry and 'other land uses', not only in cities and towns but in 'villages, hamlets and isolated developments (including farmsteads)' (Best and Anderson, 1984, p. 23).

The most recent data suggests that, between 2009 and 2014, the total amount of agricultural land actually increased by 88,000 hectares in the UK and 28,000 hectares in England (DEFRA, 2014). Best (1981) estimated that, by 2001, the proportion of urban land would go up to 10.6 per cent in Great Britain and 14.1 per cent in England and Wales. Analysis for the Foresight Land Use Futures Project (2010) suggested that his predictions were broadly accurate. Their data showed that, in 2005, 72 per cent of the land in England was in agricultural use and only 9.95 per cent was 'developed'. A further 6.9 per cent was 'Other green space', at least some of which was included in Best's definition of urban land. At the Great Britain level, 73.7 per cent of land was in agricultural use in 2005, continuing the trend shown in Table 3.1.

Some have questioned whether the protection of agricultural land is a worthwhile objective of green belts, both because 'Farmland adjacent to urban areas is notoriously difficult to farm efficiently' (Hague, 2005, p. 165) and because of the impact of policies such as the EU Common Agricultural Policy, which generated the famous "butter mountains" in the 1980s (Willis and Whitby, 1985). Neoliberal economists decry the protection of 'inefficient UK farmers from foreign competition' (Papworth, 2015, p. 19). There is perhaps little doubt that, in 2015, the UK was in a strong position to import as much food as it needed, but this of course may not be the position in perpetuity. In the recent past the UK government has begun to stress the need for greater self-sufficiency, or at least the *potential* for self-sufficiency (House of Commons, 2014), so we cannot assume that the agricultural case for green belts can be ignored. However, establishing the impacts of green belts does not necessarily require in-depth assessment of land-use statistics – one can simply observe the sharp boundary around many urban areas that delineates urban from non-urban land (see Figure 3.3).

Such 'clinical lines of land use distinction' (Hoggart, 2005, p. 5) will usually indicate the presence of the green belt, though in some cases other policies are used to similar effect. Studies that have explored the opinions of local stakeholders have found broad consensus that green belts have had this effect (DoE, 1993; Hague, 2005), while Gunn reviewed all the regional spatial strategies in place at the time and found that 'all draw particular attention to the ability of their green belts to prevent sprawl' (2007, p. 610). Most of them also mentioned the limiting of urban coalescence. Using 'allometric ... [and] geometric relations', Longley *et al.* (1992, p. 437) showed the effects of green belts on particular settlements, arguing that they had indeed limited urban growth and led to a more compact urban form. A major report produced by the CPRE and Natural England found that, between 1985 and 2006, 1 per cent of land in England had been converted from undeveloped to developed use, compared with 1.5 per cent of urban land, 0.7 per cent of rural land and 0.9 per cent of green belt land, and that 'significantly more development takes place around major towns and cities without a Green Belt than those with one' (CPRE and Natural England, 2010, p. 20). Specifically within Birmingham during that period, the development rate for green belt land was half that of the city as a whole (ibid.).

As we found in Chapter 2, much of the work discussing the impacts of green belts has focused on London. An interesting piece of work published in 2007 attempted to model how London would have developed if the pre-1939 rate of growth had been allowed to continue in an unrestricted way, without the controls put in place by the 1947 Town and Country Planning Act and the subsequent guidance on green belts. Green (2007) estimated that the 'core area' of London, in the year 2000, would have expanded to roughly 500,000 hectares in size, approximately double its actual size at that time of 275,000 hectares, and illustrated this with striking images that show the city encompassing towns such as Reading and Southend. Such towns, while currently beyond the *administrative* and *physical* boundaries of London, are quite clearly closely related in *functional* terms. It is worth reflecting on what these terms mean, as they are important in the context of how green belts have operated.

Physical vs. functional containment and leapfrogging

It is easier to illustrate different understandings of urban areas using the example of Liverpool, as London is so large that its effects stretch much further than those of other cities. The administrative boundary of Liverpool (i.e., the area covered by Liverpool City Council) is drawn quite tightly, as shown in Figure 3.1. Physically, the urban area of which Liverpool is the centre extends beyond this boundary, in some cases by quite some distance. The green belt, shown in Figure 3.2, limits the spread of Liverpool in some areas but was imposed after the

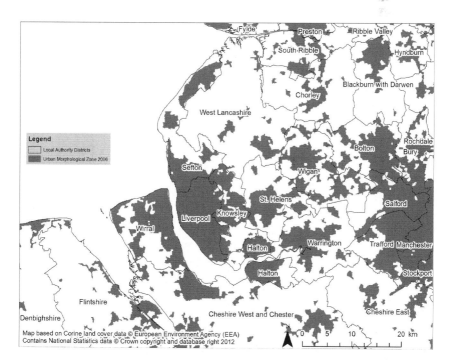

Figure 3.1 The functional urban area of Liverpool (thanks to Andreas Schulze Bäing for Figures 3.1, 3.2, 3.4 and 3.5)

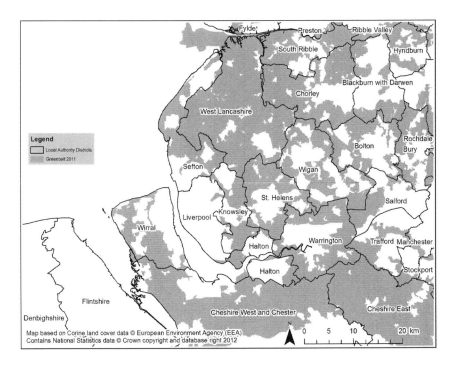

Figure 3.2 The green belt around Liverpool

city had coalesced, at least partly, with towns to its north and east (in the local authorities of Sefton, Knowsley and St Helens). Without the green belt, it seems fair to assume that much of the land immediately around Liverpool would have become urbanised. For example, Figure 3.3 shows the edge of the Speke housing estate, begun in 1937 and planned to be developed on garden city principles – indeed, it was lauded in the 1940 Barlow Commission report for being an example of good planning of housing and industry together (Barlow, 1940). The estate was completed shortly after the Second World War and at that time was intended to be 'a self-contained "community unit"' (Ravetz, 2001, p. 101), although it is now part of the urban fabric of the city. The green belt runs around its edge and, as Figure 3.3 shows, marks a clear break with that urban fabric. As noted above, we cannot simply assume that land beyond Speke would have been developed if it had not been designated as green belt, but it seems likely, given the expansion of the city in other directions. Figure 3.4 shows how little land there is around Liverpool that is currently neither urban nor green belt – so-called white land, literally white on this map. The scope for greenfield development in and around the city is severely limited.

Figure 3.5, a fourth map of the area around Liverpool, includes the Travel to Work Area (TTWA) of Liverpool, based on 2011 census data. There are 228 TTWAs covering the whole of the UK, and they show areas with a

Figure 3.3 Speke housing estate, on the outskirts of Liverpool (source: Authors' own)

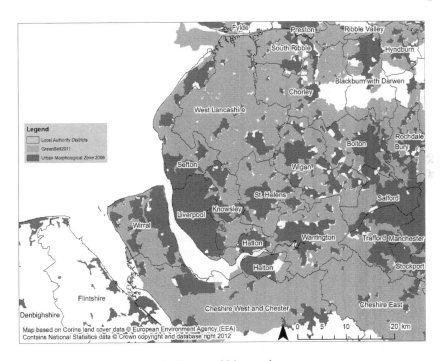

Figure 3.4 The lack of "white land" around Liverpool

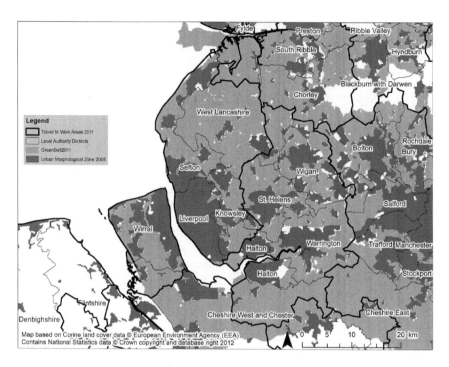

Figure 3.5 Liverpool and surrounding Travel to Work Areas (TTWAs)

66.7 per cent or higher 'self-containment rate' (ONS, 2015) – i.e., the proportion of the area's population that also works in the area and the proportion of the area's workforce that also lives in the area. As can be seen in Figure 3.5, Liverpool's TTWA includes all of the adjoining local authority of Sefton, along with most of Knowsley and West Lancashire local authorities. Many of the people within this TTWA must therefore live beyond the green belt around Liverpool. They may work in the same immediate area where they live, but it is likely that many of them commute into Liverpool. So the green belt and other planning policies have affected the *physical* containment of the city but not the *functional* containment, resulting in, among other things, longer commutes for many workers.

The risk of this unwanted impact of green belts was identified in 1962 by the MHLG, which concluded that commuting distances would increase in length 'if the outward movement of people in search of homes is not accompanied by a similar outward movement of employment' (MHLG, 1962, p. 25). Only eleven years later, Peter Hall and his colleagues, in their landmark *Containment of Urban England*, found that, although urban growth had been contained, sub-urbanisation (defined by them as residential areas moving further away from employment areas) had continued to take place. They concluded that the only effect of policies such as green belts was that the suburbs had been pushed further away from city centres than if the system had not been in place. This was caused by both the lack of employment decentralisation warned against by the MHLG in 1962 (see Chapter 2) and the rapid rise in car ownership (Hall *et al.*, 1973b). Others have continued to make the case that green belts have physically but not functionally constrained cities at intervals ever since (see, for example, Simmie, 1993; Ward, 2004), some arguing that this was the fault not of green belts but of a lack of regional planning (DoE, 1993). The most recent data on TTWAs, published in September 2015, showed that they had increased in size since 2001, leading some to argue that green belts in place around cities such as Oxford and Cambridge have led to their TTWAs growing because, 'by constraining housing growth immediately around the cities, they force commuters to drive across them to get to work, displacing rather than preventing sprawl' (Tilley, 2015, p. 15). This "displacement" of sprawl is sometimes called "leapfrogging".

As noted above, leapfrogging was identified as a concern in relation to green belts – or, rather, in relation to green belts in the absence of firm and sustained decentralisation of jobs – in the early 1960s, if not before. In the preceding sub-section we discussed these issues in the context of Liverpool, but it is in places where there is very high demand for housing (unlike Liverpool) that leapfrogging can be most prevalent. Researchers have identified leapfrogging as being a particular issue in Aberdeen (Hague, 2005), Edinburgh (DoE, 1993; Bramley *et al.*, 2004), London (Papworth, 2015, among others) and Dundee, where it was cited as one of the reasons for abandoning the green belt around that city in the 1980s (Bramley *et al.*, 2004). Gunn, in her review of regional spatial strategies, found concern was expressed in relation to the East and South West of England that green belts had caused 'development to leapfrog the greenbelt. In both cases this is seen as inherently less sustainable than a more contained development pattern' (Gunn, 2007, p. 610). We return to this issue of sustainability below, following a consideration of the impact of green belts on urban *form*.

Green belts and urban form

There seems to be a broad consensus that green belts and the broader approach of containment have indeed limited urban sprawl. This was one of the main aims of the founders of the green belt as we know it today – Patrick Abercrombie principal among them. But, as we noted in Chapter 2, Abercrombie assumed that the (urban) population, both in London and in the country more widely, would not increase, whereas the UK urban population grew from 40 million in 1950 to 53 million in 2015 – a 32 per cent increase (United Nations Department of Economic and Social Affairs, 2014). The number of households in the UK more than doubled in the same period, from 13.3 million in 1951 to 26.7 million in 2014 (Holmans, 2012; ONS, 2014). Where, beyond leapfrogging, which explains only a fairly small proportion of new development, have these new households been located? The answer, broadly speaking, is *within* existing urban areas, leading to a densification of those areas.

This was noted by some in the early 1970s (Warren-Evans, 1974), and Hall *et al.* (1973b) observed the apparent perversity of urban density intensifying as the size of households increased, where we might typically expect the reverse – i.e., that house sizes would expand in line with household size. This growing *urban intensification* is sometimes called 'town cramming' (Bramley *et al.*, 2004) and sees a redevelopment of land (sometimes green space) within urban areas (DoE, 1993; Papworth, 2015). The average size of new houses has also fallen (Cheshire and Sheppard, 1989; Evans, 1991), often as prices have increased – a topic to which we return below. In order to fit more homes into a smaller space, housebuilders have adopted various tactics, including building very close to the road to allow a reasonably large back garden, and householders have sold their own back gardens for housing (Evans, 1991), which in turn impacts on the broader urban realm.

A second consequence in relation to urban form is that the green belt policy has prevented the development of new forms of urban growth, such as new villages, towns and cities or large urban extensions (Warren-Evans, 1974; DoE, 1993), resulting in what has been called as long ago as 1965 '"the preservation of an archaic settlement pattern"' (cited in Scottish Executive Central Research Unit, 2001, p. 11). It has been argued that the 'compact city' is more sustainable (Urban Task Force, 1999), and this was the direction of planning policy in the UK for at least the period 2000 to 2010, but equally others have argued that there is little evidence for this, so 'we must infer that the policy prescriptions in favour of compact, contained cities are rooted as strongly in values as in science' (Hague, 2005, p. 162) and that the compact city position is one adopted by elitist hypocrites who often themselves live in low-density suburban areas (see Bramley *et al.*, 2004, p. 38; Papworth, 2015). In the next section we explore these arguments in more detail.

The sustainability or otherwise of green belts

As with many other terms we use in this book, we recognise that "sustainability" is inherently a contested concept, and that is before we even consider whether green belts make a positive or negative contribution to sustainable development. The way sustainable development is frequently conceptualised is to involve a balance between the three "legs" of environmental, social and economic progress and to consider the long-term as well as the short-term impacts of decisions. While the tripartite conception of sustainability has been extensively criticised, it remains central to planning decision-making in the UK, so we use it here to organise this section of the chapter.

Environmental impacts of green belts

The CPRE and Natural England (2010) categorise 66 per cent of green belts as agricultural land. It has been claimed that approximately half of this (i.e., a third of green belts overall) is in use for intensive farming, which it is further claimed generates aggregate environmental costs (Cheshire, 2014b; Papworth, 2015). CPRE & Natural England also calculated that 53 per cent of agricultural green belt land was designated under agri-environment schemes, in comparison with 67 per cent of farmland nationally, which 'suggests lower environmental quality and potential for schemes to deliver outcomes' (2010, p. 71).

Agricultural land is likely to be further out from urban areas. A number of researchers have noted that land on the immediate urban fringe is more typically associated with so-called bad-neighbour uses and/or those that require a larger land footprint, so are not economical to include in city centres, such as waste recycling, park and ride schemes, scrapyards, quarries, etc. (Bramley *et al.*, 2004; Gallent *et al.*, 2006; Gant *et al.*, 2011). While, on the face of it, the environmental value of such sites may seem limited, it is worth remembering that many researchers now argue that brownfield land has more biodiversity potential than some agricultural land (Hunter, 2014). CPRE & Natural England (2010) assessed green belt land against various environmental indicators, including biodiversity. In summary, in comparison with all land in England, green belt land performs well against some criteria and average against others. For instance, slightly more green belt land is used for forestry (11 per cent) when compared to the whole of England (8 per cent), but approximately the same proportion of green belt land is vacant or derelict brownfield land (0.2 per cent).

The positive environmental benefits of green belts internationally in terms of species preservation and mitigation of air pollution and climate change have been identified by Herath *et al.* (2015). Similarly, it has been stated that environmental protection has become increasingly important in recent years as a rationale for the green belt in Seoul and that, without it, 'Seoul would have lost much of its rich natural heritage and essential ecosystem services' (Bengston and Youn, 2006, p. 11).

So there is some evidence that green belts make a positive environmental contribution in some ways. Conversely, and beyond the potential negative impact of intensive farming, others have suggested that the pattern of land use resulting from some green belts – i.e., leapfrogging – has negative environmental impacts as well as the social problem of longer commuting journeys. As noted above, we have written elsewhere about what we consider, along with others, to be the socially regressive tendency to label housing development in rural areas as inherently unsustainable, usually on the basis that it will encourage more car use and hence increase carbon emissions (Hoggart and Henderson, 2005; Taylor, 2008; Sturzaker, 2010; Sturzaker and Shucksmith, 2011). However, if environmental considerations are part of the debate about the impacts of green belts, we must briefly consider this. Arguments have been made that, from this perspective, green belts are arguably less sustainable than alternative development patterns, particularly leapfrogging (DoE, 1993; Scottish Executive Central Research Unit, 2001; Hague, 2005). Breheny *et al.* (1993) assessed the sustainability of five forms of urban development: peripheral urban extensions, new settlements, urban infilling, key villages and multiple village extensions and found that the first was most sustainable overall, with the second and third following shortly after. The fourth and fifth, which we might see as being linked to leapfrogging, were found to be the least sustainable.

Social impacts of green belts

As noted above, Power (2001) has argued that urban sprawl causes or exacerbates various social ills in inner cities, including poverty and neighbourhood fragmentation. Ergo, following the argument above that they have limited urban sprawl, green belts should in turn have inhibited these social problems. Something Power does not explicitly discuss is the argument that urban sprawl has negative health implications, as people are less likely to walk or cycle between places. This argument is often based on evidence from the USA, where both urban sprawl and some health conditions, notably obesity and diabetes, are more widespread (Frumkin *et al.*, 2004; Guettabi and Munasib, 2014). However, some have questioned the transferability of these findings to the UK, on the basis that suburban developments in this country are more likely to include facilities within walking distance rather than people being reliant on the private car, as is the historical tendency in the USA (Papworth, 2015). Increasing opportunities for outdoor recreation, with the benefits for physical and mental health that this can bring, is often cited as something that green belts can do, both internationally (Bengston and Youn, 2006) and in the UK – indeed, it relates to two of the six objectives for green belts in the old PPG2 (DoE, 1995). It is fair to describe the evidence for this claimed impact of green belts as somewhat mixed.

A survey of some 3,000 users of eighty-three sites in the London green belt some years ago found that 'few residents of the Inner City use the Green Belt for informal recreation' (Harrison, 1983, p. 312), and that most visitors to the green belt lived in close proximity to it. Non-car owners were one group who could not make as much use of the green belt, and the fact that the proportion of UK households with no car fell from 40 per cent at the time of the study to 20 per cent in 2010 (ONS, 2010) will have some effect on accessibility of the sites Harrison surveyed. More recent work suggests that there are marked variations in the background of those using green belts – while 58 per cent of people surveyed by the CPRE and Natural England (2010) had visited the green belt within the last twelve months (including for recreation), this figure fell to 32 per cent for respondents from a black or minority ethnic background. The type of recreation enjoyed in green belts is also quite specific: while there is a higher density in green belts of public rights of way and national cycle networks than there is in non-green belt land, demonstrating 'the greater *potential* for local walks and opportunities to explore the countryside in the urban fringe' (ibid., p. 34, emphasis added), the rate of equestrian facilities per 1,000 households is fifteen times higher within the green belt than in England as a whole. Papworth (2015) suggests that people living more than about a kilometre from green belts are much less likely to use them for recreation, which supports another of Harrison's findings – that green belts 'have more in common with the use of urban open space than with day-trips to the countryside' (Harrison, 1983, p. 312).

The desire to access the green belt for its recreational value is one of the basket of 'amenity values' that are captured by econometric models that seek to value the green belt. One of these, *hedonic price modelling*, assesses the positive value of proximity to the green belt on house prices (this is returned to below), but another type of modelling, *contingent valuation*, involves asking people about their will-ingness to pay (WTP), or willingness to accept compensation, in a hypothetical market for an environmental good such as the green belt. There are a number of studies that make use of contingent valuation to estimate the value of green belts.

One such study looked at the historical city of Chester in the North West of England and at arguments around its 1990 Structure (strategic) Plan, which pro-posed green belt deletions of 417 hectares for 7,800 new houses, and specifically attempted to value the costs for one of the de-designation sites of 38 hectares, for 900 houses, via a hypothetical trust fund to buy the site. 54 per cent of respondents indicated a willingness to pay a positive amount to secure the site from develop-ment, at a mean WTP of £43 per individual. Local estate agents estimated the value of the site for housebuilding at £2.268 million, so, if a crude calculation was done by dividing this by the population of Chester, each individual would have to pay £24.23 to preserve the land. If respondents were bidding for the whole 417 hectares of green belt proposed for deletion (maps of which were shown to them), their bids were insufficient (Hanley and Knight, 1992). An earlier study of the Newcastle green belt by Willis and Whitby (1985) came up with similar findings – that the green belt had an amenity value for the population of the city but that one of a considerably smaller size would deliver a similar amenity value. More recently, Barker (2004) cited research by the then Office of the Deputy Prime Minister that showed a much lower value placed on green belts than on city parks – the latter being valued at an extraordinary sixty times more than the former.

Finally, Hague (2005) has argued that the Edinburgh green belt in particular, and green belts more broadly, have had negative impacts on social cohesion and social justice, through separating homes and workplaces and through the socially regressive effects of green belts. These are discussed in the next subsection and returned to in Chapter 7.

Economic impacts of green belts

As noted above, Peter Hall and his colleagues had identified in the early 1970s that green belts had caused a separation between workplace and home (Hall *et al.*, 1973a, 1973b). They went further, arguing that this separation was regres-sive, as those who were most disadvantaged by this were those without cars and/ or those in 'the lower-income levels of the home buying market' (Hall *et al.*, 1973b, p. 407), who had bought, or aspired to buy, houses since the implemen-tation of 'the Containment of Urban England'. Conversely, those who made the greatest gains were those who owned land and property at the outer edge of the built-up area and in the countryside, as they benefit the most from house price rises (see below) and have access to an undisturbed countryside. Others shared this analysis (Evans, 1973; Simmie, 1974) at the time, and similar arguments have been made at intervals since (Simmie, 1993; Cheshire and Sheppard, 2002; Ward, 2004; Papworth, 2015); both Hall and Simmie described the result as residential 'apartheid'. Much of this criticism relates to the increased cost of housing caused by green belts, but some considers specifically access to green space, whether for recreation or other purposes. Cheshire and Sheppard (2002) and Papworth (2015) argue that preserving green belts is regressive because, in the words of Papworth, they protect 'large amounts of plentiful green space around rich people at the expense of rare green space near poorer people' (ibid., p. 32). In the American context, the green belt around Boulder, Colorado, and the wider landscape protection approach of which it is part, has been described as both regressive and exclusionary – a tool to maintain the "whiteness" of Boulder (Hickcox, 2009).

Other arguable economic impacts of green belts relate to their potential to assist inner-city regeneration (a purpose of green belts in England since 1984) and, conversely, the risk they limit economic growth. In terms of regeneration, Barker argued that, in some parts of the country, including the North West, green belts had played 'a vital role in securing regeneration' (2006, p. 62), and Gunn (2007) found that regional spatial strategies in the North and the West Midlands made similar arguments. However, others (Elson, 1986; DoE, 1993; Papworth, 2015) have pointed out that, as with other impacts of the green belt, there is little direct evidence to support this assertion. The assumption that constraining urban growth means that economic development will occur within the city centre, as opposed to on 'sites with similar attributes in other parts of the outer city' (DoE, 1993, p. 21), is unproven and perhaps dubious. In terms of economic development more broadly, the argument is made by some (Evans and Hartwich, 2007; Papworth, 2015) that the green belt increases (land and transport) costs for businesses, which in turn raises prices for households and, *in extremis*, can lead to businesses relocating overseas, where there are fewer constraints on development. The logic here is clear enough – that businesses want larger sites with more parking, perhaps nearer to the main road network (Hague, 2005), and if land costs are higher they will either relocate or be unable to expand (Gant *et al.*, 2011). However, there is also evidence that local authorities operate a "looser" approach to green belts where they are keen to support economic growth (DoE, 1993), so again the evidence is mixed.

Finally in this chapter we move on to the topic which receives the most attention in terms of the impacts of green belts – that of land and house prices.

Impacts on land and house prices

It is important here to reiterate how land is developed for housing in the UK, and specifically England, as it is on the latter nation that much attention has focused in relation to green belts and house prices. During the period 2004 to 2010, the need/demand for housing was calculated at the regional level, and figures of housing numbers to be provided were set out in regional spatial strategies (RSSs) divided between local planning authorities (LPAs). Each LPA was then required to plan for that number of houses, with the vast majority to be built by the private sector (see Chapter 1). The 2010–2015 coalition government abolished regional spatial strategies and consequently the setting of regional targets. LPAs are now 'encouraged to calculate their own housing figures and set aside enough land to satisfy housing demand. Although not mandatory, LPAs are encouraged by Government to have a Local Plan adopted … which sets out housing need in the particular area' (Smith, 2015, p. 3). Once they have calculated their housing need, drawing upon government estimates of household growth, demographic trends, etc., LPAs may allocate land for housing in their local plan to meet this need (sometimes referred to as "releasing land"), or they may rely on "windfall" sites – i.e., sites that are brought forward by a developer for housing independently of the local plan. Either way, they are required to 'identify and update annually a supply of specific deliverable sites sufficient to provide five years' worth of housing against their housing requirements with an additional buffer of 5%' (DCLG, 2012, p. 12). In October 2014 the government issued guidance to LPAs to (re)affirm that housing need *did not* outweigh the protection given to green belts, and that 'Green Belt boundaries should only be altered in exceptional circumstances' (DCLG, 2014). It is clear, then, that an LPA allocating land for housing in its local plan, or giving it planning permission for housing, greatly increases the chances that housing will be built on it, and consequently increases the value of that land (agricultural land in the UK being worth substantially less than housing land).

One simple way to establish the impact of green belts, and other forms of housing restraint, would be to calculate the difference in price between agricultural and housing land. However, this difference in price, although a cost to those who buy housing land and build houses on it, and subsequently of course those who buy those houses, is a benefit to someone (the landowner) – so from one perspective is simply the market in operation. Others, however, argue that

> The economic cost of the increase in land prices derives from the fact that space is something from which people derive welfare or benefits … Thus the economic cost that is paid for the planning system … is the reduction in welfare that occurs because higher land prices entail lower consumption of housing and land.
>
> (Cheshire and Sheppard, 1989, pp. 472–3)

There is also an interesting conceptual debate about whether high land prices cause high house prices or vice versa, which depends in part on whether one believes that high demand or constrained supply causes the price increase. Cheshire and Sheppard argued that

> there is joint causation … If development control effectively restricts the supply of residential development then it raises the prices of houses. Since builders as a result can get higher prices for houses, they will pay more for land. While, therefore, the high land prices do not cause high house prices, both are caused by the restriction of development.
>
> (Ibid., p. 471)

As noted above, one area of research relating to the costs of green belts is that of *hedonic price modelling*, which correlates variations in house price with proximity to what are perceived to be assets or liabilities – public transport, parks, water, airports, etc. – the theory being that the benefit (or disbenefit) of these features is 'internalised' into housing and land prices (Nelson, 1985). There is a body of work which applies this approach to green belts in order to establish either a notional "value" to green belts in amenity terms, seeing increased house prices as a *positive* effect of green belts (i.e., looking at the demand side), or the impact of the green belt on limiting housing supply, and hence increasing house prices as a *negative* effect (i.e., looking at the supply side).

Hedonic modelling

International studies

Herath *et al.* (2015) apply a hedonic approach in Vienna and assess the impact of proximity to the green belt, the central business district and the *Prater* (an amusement park and green space) on house prices. Their conclusion is that the green belt has a 'significant impact on apartment prices' (2015, p. 369), more so than the other two factors. There are several studies looking at hedonic modelling in Seoul. Bengston and Youn review a number of them and conclude that, overall, the Seoul green belt shows a 'relatively modest' (2006, p. 10) effect on house and land prices, with other supply (and demand) factors being equally important. Lee and Linneman (1998) do something comparatively unusual and assess the effect of the Seoul green belt longitudinally – i.e., they look at how it has changed over time, from 1972 to 1989. Their conclusion was that the effect had diminished over this period – in 1972, there was a 75 per cent negative influence on residential land values within the green belt (i.e., a strong assumption that development was unlikely in this area), which by 1989 had fallen to 40 per cent – reflecting the belief that, as the area inside the green belt was developed to capacity, the local government would release some of its land for development. The effect of accessibility to the green belt in amenity terms had also reduced over that period – there was a 4 per cent fall in value with every kilometre distance from the green belt in 1989, down from a high of 18 per cent in 1980. Lee and Linneman ascribed this fall in amenity value to the notion that the green belt is a 'congestible local public good' (1998, p. 109) – i.e., the amenity value is high initially when the level of use is small, but, as it becomes more popular, the value that each person gains from it falls.

There are also a number of studies looking at the impact of green belts and similar policies in the USA. In relation to the green belt around Boulder, Colorado, it has been estimated that, all else being equal, 'the average value of properties adjacent to the greenbelt would be 32% higher than those 3,200 walking feet away' (Correll *et al.*, 1978, p. 211) – with a gradation showing, at that time, 'a $4.20 [approximately $15 in 2016] decrease in the price of a residential property for every foot one moves away from the greenbelt' (ibid.). The effect of the urban growth boundary (UGB) in Portland, Oregon, is the focus of various studies. It is important to note that this is not a green belt but, rather, a boundary denoting that, beyond it, urban development will not be allowed. There is thus a much reduced amenity value to being adjacent to the green belt, so studies of Portland look at the impact on house prices from a negative – i.e., constraint – perspective, with mixed results. Knaap (1985) found that the UGB, implemented in 1980, had had a significant effect on land, and consequently housing, values at that time. Later studies (Dawkins and Nelson, 2002; Jun, 2006) conversely found that, over the longer term, the Portland UGB, and others in Oregon, had *not* led to an increase in house prices because, at the regional level, 'policymakers were particularly keen to incorporate additional policy levers to ensure than urban containment does not constrain the supply of land for housing and economic growth' (Dawkins and Nelson, 2002, p. 7). This is in contrast to those in California, where containment policies were often specifically linked to limiting housing output – the end result being in that case that land and house prices went up. Dawkins and Nelson, then, made the case that factors other than the approach to planning at the city level were equally, if not more, influential on house prices. We return to this argument below.

Hedonic studies in the UK

There is of course always a difficulty in relating findings from studies outside the UK to the planning system and housing market in this country on account of the context-specific nature of such things. For instance, the Oregon studies cited above took place in a context where a regional (state) level policy was in place to manage issues of urban growth at the strategic level above that of cities. In the UK, or at least in England, such a policy has come in and out of existence since the 1950s and is not currently in place (in 2016). But there are some notable examples of hedonic price modelling in England in relation to the green belt and the broader regime of urban containment. As is the case elsewhere, some studies look at this from a demand/amenity perspective, some from a supply/constraint perspective.

A study of the effect of London's green belt on house prices in 1968 estimated a 4.9 per cent increase for houses within 2 miles of the green belt, resulting in an amenity value of each hectare of the London green belt of £2,944 [equivalent to around £46,000 in 2016] (cited in Willis and Whitby, 1985). Gibbons *et al.* (2011) looked at a range of different features and modelled their impact on house prices in amenity terms. They found that being in a National Park added an average of £9,400, and being within a green belt £5,800, to house prices for transactions within major metropolitan areas.

Cheshire and Sheppard (1989) compared planning regimes and house prices in Darlington (a town in the relatively deprived north east of England with a relaxed planning regime and low house prices) and Reading (a town in the relatively wealthy south east of England with a strict/constrained planning regime and high house prices). They then theoretically applied the planning regime as it is in Darlington to Reading, both "internally" and in terms of easing of containment policies. This led to a modelled fall in house prices of between 6 and 17 per cent at 4,000 feet from the centre of town, down to between 2 and 6 per cent 18,000 feet from the centre of town, depending on house size/type. The impact of the planning system seems to have explained, typically, about half the difference in price between a 4 bedroomed detached house in Reading compared to Darlington but only about a tenth of the difference in price for a 2 bedroomed terraced house ... people living in larger detached houses would, financially, have most to lose if planning restrictions were relaxed.

> (Cheshire and Sheppard, 1989, p. 484)

At the "bottom" end of the market, the fall in house prices and consequent reduction in the size of a mortgage required to purchase a property would have opened up the market to an additional 4 per cent of the population. But the authors estimated that the greater availability of land would lead to plot sizes growing by 65 per cent, so the area of Reading would increase by 50 per cent – probably a politically unlikely outcome.

One problem with hedonic modelling is that it is "data hungry" – it requires very large datasets to be accurate. One of the larger studies was carried out by Hilber and Vermeulen (2010) for the DCLG. They use data over a long period of time (thirty-five years) and from most, if not all, local planning authorities in England. Hilber and Vermeulen address the impact of 'regulatory restrictiveness' rather than specifically of the green belt, and claim that,

> On the basis of this rich dataset, this report provides unambiguous *causal* evidence demonstrating that regulatory supply constraints ... have had a serious negative long-run impact on housing affordability and have increased house price volatility ... the increase in real house prices between 1974 and 2008 can – to a large extent – be explained by the existence of tight planning constraints.
>
> (2010, p. 7, emphasis in original)

The authors go on to model what would happen to house prices if the planning system were relaxed and predict a 15 to 52 per cent fall, depending on the extent to which this were to take place. They do acknowledge that the planning system does not fully explain house price volatility, as other factors also play a role. These other factors provide one reason why some prefer to use other, non-hedonic, models to estimate the impact of the green belt on land and house prices.

Other approaches to modelling the impact of green belts on land and house prices

Several studies have used data comparing the much higher rate of house price inflation to other measures of inflation to argue that, in straightforward economic terms, constraining the supply of housing does, indeed must, lead to a rise in prices. Alan Evans (1991), building on his previous work on the topic, called attention to the fact that, between 1969 and 1989, the retail price index in the UK went up eight times and land prices went up sixty times – a point he expanded on several occasions (Evans, 1973, 1985, 1987, 1988, 1991, 2004). He argued that there was money to be made – and hence businesses that specialised – in obtaining planning permission for housing on agricultural land and then selling it on at a substantially increased value. An economist would call this "rent-seeking behaviour", and since it 'does not in any way increase economic welfare … it is a dead-weight loss' (Evans, 1985, p. 206). No gain to society is made by such conduct, and Evans's argument is that it illustrates a cost to society of the planning system.

Another prominent economist who has come up with similar opinions is Paul Cheshire (see above for a hedonic study). He and colleagues (Cheshire *et al.*, 1999) estimated the impact of a target then recently put in place by the (Labour) government to see 60 per cent of all new housing built on brownfield land. They calculated that this would increase house prices by 132 per cent by 2016, because 'the overwhelmingly more important driver of the demand for "housing" was not household numbers: it was rising real incomes' (Cheshire, 2009, p. 12). Having modelled house price change without a change in income, they found that it would increase by only 4.4 per cent. As noted above, income change is not something that LPAs consider when establishing their housing figures – these are based on projections of population change. Because 'income growth has been a bit lower than we assumed' (Cheshire, 2014a), the rise in house prices was a little lower than they estimated, but still very high. Cheshire concludes, therefore,

> that to stabilise, even to slow the rate of increase of un-affordability … we have to build on Greenfield sites. That necessarily means parts of the Greenbelt particularly, since the strongest demand for housing is near to jobs and near to where houses are the most expensive.
>
> (Cheshire, 2014c)

A major study on behalf of the Department of the Environment (Gerald Eve, 1992) looked at four case studies on a continuum of planning constraint, from Reigate (a town 20 miles south of London in the county of Surrey) as the most constrained, through Wokingham (a town 40 miles west of London in Berkshire), Beverley (a reasonably affluent town near the city of Hull) and Barnsley (a relatively deprived town in Yorkshire) as the least constrained. This study assessed the effect of planning constraints by comparing land prices in Barnsley (the cheapest) with those in the other three places and calculated that, in 1990, land price constituted between 16 and 42 per cent of house prices in Beverley, Reigate and Wokingham, and that this proportion had gone up between 1975 and 1990. The authors' conclusions were that land supply does affect house prices in the medium and long term and that planning has a significant effect on land supply. However, they also found that the greater responsiveness of the planning system in the 1980s (i.e., a relaxation in constraints) did not reduce prices, because the increase in demand was greater than the increase in supply. So, 'it would require significant additional land release on a national scale, and over a considerable time period, to have a measurable effect on house prices' (ibid., p. 51). Another important finding of this study, and a subsequent one (Joseph Rowntree Foundation, 1994), was that *substitutability* was limited – i.e., relaxing supply in one part of the country could not, to a large degree, compensate for constraint in another part. Both studies also identified that the prices developers were paying for sites would not lead to a profit unless the price of houses increased, suggesting they were anticipating continued inflation – and, potentially, contributing to it (Monk *et al.*, 1996). In a smaller and more recent study, Whitehead *et al.* (2015) approached the issue from the opposite perspective, that of homeowners who might oppose new residential development in their neighbourhood partly because they feared it would reduce the price of their own house. Their conclusion was that any such reduction was limited and short term. This study is interesting primarily because it was funded by the housebuilding industry – presumably to contribute to the reduction in opposition to new housing – but it makes the opposite point to a recent report by the Home Builders Federation (2014), which argues that more housebuilding is needed to reduce prices.

Glenn Bramley (1993a, 1993b) developed a large model that reflected the complexities of the supply side of the housing market and included variables such as demography, economic factors, planning policy, geographical differences, housing supply and house prices. He used data from ninety districts within twelve counties across the South East, South West, East Midlands and West Midlands of England, so excluding London and the North but including areas of high and low demand, meaning he could account for differences between regional and county-level policy. He modelled 'what would happen if the government required local planning authorities to double the amount of land to be released over the ten-year plan period, 1986–96' (Bramley, 1993b, p. 1037), firstly if they did so without any green belt deletions. He found that the increased number of houses built would not reflect the amount of land released, so housing supply might go up by 28 per cent at most because, of course, private developers, not planners, control when houses are built – what Bramley calls the 'implementation gap'. Similarly, the reduction in house prices is, at its highest, 11.5 per cent, tailing off quickly shortly after initial land releases. Bramley then includes green belt deletions in his calculations, at the same rate as building on unconstrained land, and concludes there would be marked geographical differences, with large increases in building in the West Midlands, for example, where green belt constraints have been particularly strong. However, the marginal effect of including green belt, overall, is only a maximum 3 per cent increase in housebuilding, falling to 2 per cent on average, and a 1.2 per cent effect on prices. So 'the benefits of extending a land-release strategy into the greenbelt are very marginal, and would surely be outweighed by the political and environmental costs' (Bramley *et al.*, 1995, p. 157). Evans (1996) felt that Bramley's model would understate changes in house prices to at least some degree and has argued previously (see above) that planning *does* cause an increase in house prices.

This section has presented a range of evidence from different sources about the impact of green belts on house prices. It is fair to say there is a substantial difference of opinion, with some researchers convinced that green belts *have* caused prices to be much higher than they would otherwise be, and others equally convinced that any such effect is small and that the benefits of green belts outweigh these costs to society.

Conclusion

As with the preceding section, so this chapter as a whole: there is a great deal of debate about the impacts of green belts from many commentators, and there is very little agreement between them. It is our opinion, having weighed up the available evidence, that green belts have succeeded in limiting urban sprawl – indeed, this is one topic around which there is a broad consensus, though there are differences of opinion as to the extent of leapfrogging. The evidence around house prices is extremely mixed, with strong opinions, and plentiful evidence, on both sides. We are not economists, so we cannot comment in great detail on the varying methodologies employed, but what this difference of opinion illustrates is that there are no "facts" in relation to the impacts of green belts; one must examine data and its interpretation and assess it according to one's own political and professional position.

In terms of the social, environmental and economic sustainability of green belts – again, opinions differ. Weighing up, for example, the environmental benefits that come from the land protected from development against the increased costs of commuting is not straightforward. On a utilitarian basis, it is probably impossible to say whether the aggregate costs of green belts outweigh the benefits. Our own position, as broadly left of centre in political terms, is that the costs and benefits of green belts appear to have been unequally distributed. As famously argued by Hall *et al.* (1973b), the costs have fallen on the less well off, and the benefits have been accrued by the wealthier, property-owning groups in society. The position has not changed in the forty years since the publication of *The Containment of Urban England* – indeed, the regressive nature of this 'containment' has arguably become more profound. It was this, at root, which prompted the writing of this book, and it is this which prompts us, in Chapter 6, to consider alternatives to the green belt that might match its laudable environmental achievements with a less regressive distribution of costs.

References

Abercrombie, P. (1945) *Greater London Plan* (London: HMSO).

Barker, K. (2004) *Delivering Stability: Securing our Future Housing Needs* (London: HMSO).

Barker, K. (2006) *Barker Review of Land Use Planning: Final Report – Recommendations* (London: HMSO).

Barlow, Sir M. (1940) *Report [of the] Royal Commission on the Distribution of the Industrial Population* (London: HMSO).

Bengston, D. N., and Youn, Y. C. (2006) Urban containment policies and the protection of natural areas: the case of Seoul's greenbelt, *Ecology and Society*, 11(1), www.ecologyandsociety.org/vol11/iss1/art3/.

Best, R. (1981) *Land Use and Living Space* (London: Methuen).

Best, R., and Anderson, M. (1984) Land-use structure and change in Britain, 1971 to 1981, *The Planner*, 70(11): 21–4.

Bramley, G. (1993a) The impact of land use planning and tax subsidies on the supply and price of housing in Britain, *Urban Studies*, 30(1): 5–30.

Bramley, G. (1993b) Land-use planning and the housing market in Britain: the impact on housebuilding and house prices, *Environment and Planning A*, 25(7): 1021–51.

Bramley, G., Bartlett, W., and Lambert, C. (1995) *Planning, the Market and Private House-Building: The Local Supply Response* (London: UCL Press).

Bramley, G., Hague, C., Kirk, K., Prior, A., Raemaekers, J., Smith, H., Robinson, A., and Bushnell, R. (2004) *Review of Green Belt Policy in Scotland* (Edinburgh: Scottish Executive Social Research).

Breheny, M., Gent, T., and Lock, D. (1993) *Alternative Development Patterns: New Settlements* (London: HMSO).

Cheshire, P. (2009) *Urban Containment, Housing Affordability and Price Stability – Irreconcilable Goals* (London: Spatial Economics Research Centre).

Cheshire, P. (2014a) Building on greenbelt land: so where?, http://spatial-economics. blogspot.co.uk/2014/07/building-on-greenbelt-land-so-where.html.

Cheshire, P. (2014b) Turning houses into gold: the failure of British planning, *CentrePiece (the Magazine of the Centre for Economic Performance)*, 19(1): 14–18.

Cheshire, P. (2014c) Where should we build on the greenbelt?, http://blogs.lse.ac.uk/ politicsandpolicy/building-on-greenbelt-land/.

Cheshire, P., and Sheppard, S. (1989) British planning policy and access to housing: some empirical estimates, *Urban Studies*, 26(5): 469–85.

Cheshire, P., and Sheppard, S. (2002) The welfare economics of land use planning, *Journal of Urban Economics*, 52(2): 242–69.

Cheshire, P., Marley, I., and Sheppard, S. (1999) *Development of a Microsimulation Model for Analysing the Effects of the Planning System Housing Choices: Final Report* (London: London School of Economics, Department of Geography and Environment).

Coleman, A. (1977) Land-use planning: success or failure?, *Architects Journal*, 165(3): 94–134.

Correll, M. R., Lillydahl, J. H., and Singell, L. D. (1978) The effects of greenbelts on residential property values: some findings on the political economy of open space, *Land Economics*, 54(2): 207–17.

CPRE and Natural England (2010) *Green Belts: A Greener Future* (London: CPRE & Natural England).

Dawkins, C. J., and Nelson, A. C. (2002) Urban containment policies and housing prices: an international comparison with implications for future research, *Land Use Policy*, 19(1): 1–12.

DCLG (2012) *National Planning Policy Framework* (London: Department for Communities and Local Government).

DCLG (2014) Planning practice guidance, 9: Protecting green belt land, http://planningguidance. communities.gov.uk/blog/policy/achieving-sustainable-development/delivering-sustainable-development/9-protecting-green-belt-land/.

DEFRA (2014) Structure of the agricultural industry in England and the UK at June, https:// www.gov.uk/government/statistical-data-sets/structure-of-the-agricultural-industry-in-england-and-the-uk-at-june.

DoE (1993) *The Effectiveness of Green Belts* (London: HMSO).

DoE (1995) *Planning Policy Guidance 2: Green Belts* (London: HMSO & Department of the Environment).

Elson, M. J. (1986) *Green Belts: Conflict Mediation in the Urban Fringe* (London: Heinemann).

Evans, A. (1973) *The Economics of Residential Location* (London: Macmillan).

Evans, A. (1985) *Urban Economics: An Introduction* (Oxford: Blackwell).

Evans, A. (1987) *House Prices and Land Prices in the South East: A Review* (London: House Builders Federation).

Evans, A. (1988) *No Room! No Room! The Costs of the British Town and County Planning System*, Occasional Paper No. 79, Institute of Economic Affairs, London.

Evans, A. (1991) 'Rabbit hutches on postage stamps': planning, development and political economy, *Urban Studies*, 28(6): 853–70.

Evans, A. (1996) The impact of land use planning and tax subsidies on the supply and price of housing in Britain: a comment, *Urban Studies*, 33(3): 581–5.

Evans, A. (2004) *Economics and Land Use Planning* (Oxford: Blackwell).

Evans, A., and Hartwich, O. (2007) *The Best Laid Plans: How Planning Prevents Economic Growth* (London: Policy Exchange).

Foresight Land Use Futures Project (2010) *Final Report* (London: Government Office for Science).

Frumkin, H., Frank, L. D., and Jackson, R. (2004) *Urban Sprawl and Public Health: Designing, Planning, and Building for Healthy Communities* (Washington, DC: Island Press).

Gallent, N., Andersson, J., and Bianconi, M. (2006) *Planning on the Edge: The Context for Planning at the Rural–Urban Fringe* (London: Routledge).

Gant, R. L., Robinson, G. M., and Fazal, S. (2011) Land-use change in the 'edgelands': policies and pressures in London's rural–urban fringe, *Land Use Policy*, 28: 266–79.

Gerald Eve (1992) *The Relationship between House Prices and Land Supply* (London: HMSO).

Gibbons, S., Mourato, S., and Resende, G. (2011) *The Amenity Value of English Nature: A Hedonic Price Approach* (London: Spatial Economics Research Centre).

Green, N. (2007) What if there had been no 1947 Act or comparable measure to restrain development in the UK?, *Town and Country Planning*, 76(6): 193–5.

Guettabi, M., and Munasib, A. (2014) Urban sprawl, obesogenic environment, and child weight, *Journal of Regional Science*, 54(3): 378–401.

Gunn, S. C. (2007) Green belts: a review of the regions' responses to a changing housing agenda, *Journal of Environmental Planning and Management*, 50(5): 595–616.

Hague, C. (2005) Identity, sustainability and settlement patterns, in C. Hague and P. Jenkins (eds), *Place Identity, Participation and Planning* (Abingdon: Routledge), pp. 159–82.

Hall, P., Gracey, H., Drewitt, R., and Thomas, R. (1973a) *The Containment of Urban England*, Vol. 1 (London: Allen & Unwin).

Hall, P., Gracey, H., Drewitt, R., and Thomas, R. (1973b) *The Containment of Urban England*, Vol. 2 (London: Allen & Unwin).

Hanley, N., and Knight, J. (1992) Valuing the environment: recent UK experience and an application to green belt land, *Journal of Environmental Planning and Management*, 35(2): 145–60.

Harrison, C. (1983) Countryside recreation and London's urban fringe, *Transactions of the Institute of British Geographers*, 8(3): 295–313.

Herath, S., Choumert, J., and Maier, G. (2015) The value of the greenbelt in Vienna: a spatial hedonic analysis, *Annals of Regional Science*, 54(2): 349–74.

Hickcox, A. (2009) Green belt, white city: race and the natural landscape in Boulder, Colorado, *Discourse*, 29(2): 236–59.

Hilber, C., and Vermeulen, W. (2010) *The Impact of Restricting Housing Supply on House Prices and Affordability* (London: HMSO).

Hoggart, K. (2005) City hinterlands in European space, in K. Hoggart (ed.), *City's Hinterland: Dynamism and Divergence in Europe's Peri-Urban Territories* (Aldershot: Ashgate), pp. 1–18.

Hoggart, K., and Henderson, S. (2005) Excluding exceptions: housing non-affordability and the oppression of environmental sustainability?, *Journal of Rural Studies*, 21: 181–96.

Holmans, A. E. (2012) *Household Projections in England: Their History and Uses* (Cambridge: Cambridge Centre for Housing and Planning Research).

Home Builders Federation (2014) *Barker Review a Decade On* (London: HBF).

House of Commons, Environment, Food and Rural Affairs Committee (2014) *Food Security* (London: HMSO).

Hunter, P. (2014) Brown is the new green: brownfield sites often harbour a surprisingly large amount of biodiversity, *EMBO Reports*, 15(12): 1238–42.

Joseph Rowntree Foundation (1994) *Inquiry into Planning for Housing* (York: Joseph Rowntree Foundation).

Jun, M.-J. (2006) The effects of Portland's urban growth boundary on housing prices, *Journal of the American Planning Association*, 72(2): 239–43.

Knaap, G. J. (1985) The price effects of urban growth boundaries in metropolitan Portland, Oregon, *Land Economics*, 61(1): 26–35.

Lee, C. M., and Linneman, P. (1998) Dynamics of the greenbelt amenity effect on the land market: the case of Seoul's greenbelt, *Real Estate Economics*, 26(1): 107–29.

Longley, P., Batty, M., Shepherd, J., and Sadler, G. (1992) Do green belts change the shape of urban areas? A preliminary analysis of the settlement geography of South East England, *Regional Studies*, 26(5): 437–52.

MHLG (1956) *Report of the Ministry of Housing and Local Government for the year 1955* (London: HMSO).

MHLG (1962) *The Green Belts* (London: HMSO).

Monk, S., Pearce, B. J., and Whitehead, C. M. E. (1996) Land-use planning, land supply, and house prices, *Environment and Planning A*, 28(3): 495–511.

Nelson, A. C. (1985) A unifying view of greenbelt influences on regional land values and implications for regional planning policy, *Growth and Change*, 16(2): 43–8.

ONS (2010) Transport, in *Social Trends 40*, www.ons.gov.uk/ons/rel/social-trends-rd/social-trends/social-trends-40/social-trends-40—transport-chapter.pdf.

ONS (2014) *Families and Households, 2014: Statistical Bulletin* (London: Office for National Statistics).

ONS (2015) Travel to Work Areas, www.ons.gov.uk/ons/guide-method/geography/beginner-s-guide/other/travel-to-work-areas/index.html.

Papworth, T. (2015) *The Green Noose: An Analysis of Green Belts and Proposals for Reform* (London, ASI (Research)).

Power, A. (2001) Social exclusion and urban sprawl: is the rescue of cities possible?, *Regional Studies*, 35(8): 731–42.

Ravetz, A. (2001) *Council Housing and Culture: The History of a Social Experiment* (London: Routledge).

Reith, J. C. W. (1946) *Final Report of the New Towns Committee* (London: HMSO).

Scott, L. F. (1942) *Report of the Committee on Land Utilisation in Rural Areas* (London: HMSO).

Scottish Executive Central Research Unit (2001) *The Role of the Planning System in the Provision of Housing* (Edinburgh: The Stationery Office).

Simmie, J. M. (1974) *Citizens in Conflict: The Sociology of Town Planning* (London: Hutchinson Educational).

Simmie, J. M. (1993) *Planning at the Crossroads* (London: UCL Press).

Smith, L. (2015) *Planning for Housing*, House of Commons Library Briefing Paper no. 03741 (London: House of Commons).

Sturzaker, J. (2010) The exercise of power to limit the development of new housing in the English countryside, *Environment and Planning A*, 42(4): 1001–16.

Sturzaker, J., and Shucksmith, M. (2011) Planning for housing in rural England: discursive power and spatial exclusion, *Town Planning Review*, 82(2): 169–93.

Taylor, M. (2008) *Living Working Countryside: The Taylor Review of Rural Economy and Affordable Housing* (Wetherby: CLG).

Tilley, J. (2015) Remote working, *Planning*, 25 September, pp. 14–15.

United Nations Department of Economic and Social Affairs (2014) *World Urbanization Prospects* (New York: United Nations).

Urban Task Force (1999) *Towards an Urban Renaissance* (London: HMSO).

Ward, S. K. (2004) *Planning and Urban Change* (2nd ed., London: Sage).

Warren-Evans, J. R. (1974) The growth of urban areas, in College of Estate Management (ed.), *The Future of the Green Belt*, Occasional Papers in Estate Management, no. 5 (Reading: College of Estate Management), pp. 19–24.

Whitehead, C., Sagor, E., Edge, A., and Walker, B. (2015) *Understanding the Local Impact of New Residential Development: A Pilot Study* (London: LSE).

Willis, K. G., and Whitby, M. C. (1985) The value of green belt land, *Journal of Rural Studies*, 1(2): 147–62.

4   Characteristics of the UK green belt

Introduction

One of the main arguments we present throughout this book is the need to rethink our understanding of *what green belts are*, *what they are comprised of*, and *whether their designation is still fit for purpose*. All of these influence how our socio-economic perceptions frame these locations as well as our subsequent valuations of them. Establishing a deeper knowledge of how these factors interact provides us with a better understanding of how our views may have become entrenched in an outmoded ideal. Using Cloke's (2006) notion that green belt and rural areas should be considered not as a static concept but as a physically and socially evolving set of resources, spatial linkages and experiences (which, as we shall discuss later in the book, mirrors the thoughts of the 1940 Barlow Commission on the planning system more broadly), this chapter addresses the pluralism associated with green belt locations, our perceptions of them, and the influence that people have on their management and designations (Halfacree, 2006).

We reflect here on these issues and present a discussion of how spatial, perceptual and social-economic interpretations of green belts have shaped their use since 1947 (Cullingworth *et al.*, 2015). This will challenge any narrative that green belts should be viewed *only* as an essential conservation/preservation policy. We propose the alternative view that the values attributed to green belts are deeply problematic and often fail to acknowledge the changes in society, land-use and planning practice we have witnessed since the 1940s. Moreover, discussions of green belt have become increasingly toxic over time, in part because of the growing number of housing developments being proposed on green belt sites, which limit the opportunities to engage in a rational dialogue over their utility (CPRE and Natural England, 2010). To examine how the characteristics of green belts have influenced these debates, the chapter will be structured in three parts.

Firstly, it reflects on the *size*, *location* and *physical landscapes* that characterise green belts to examine what proportion of the country is actually covered by the designation. This includes commentary on the variation witnessed across the UK and contrasts the ways in which people visualise green belts with their actual spatial distribution and the influence this has on our reporting of their value. It also questions the perceived ecological values of green belts and discusses whether there is a disparity between the reality of land-use classifications and perception.

Secondly, our understanding of green belts is examined to question whether the *value* and *quality* arguments attached to them remain valid in a twenty-first-century context. This section examines whether the perception of green belts as being located in areas of high-quality or unique landscape character is correct and so whether this remains a valid reason to limit development in such areas. Within this argument, one of the historical contestations of rural areas in general, and green belt specifically, has been that they protect valuable landscapes from development (Macnaghten and Urry, 1998). Yet a review of their locations, their size and localised land-use characteristics suggests that this may be a flawed argument, as many green belts are located in areas that could be thought of as having marginal landscape value (Selman, 2006).

The third section reflects on the figurative, and in some cases literal, distance between the actual size and quality of green belts and the perceived interpretations of them. While the two previous sections discuss the mechanics of location and size, here we delve deeper into a more interpretive evaluation of green belts. One of the key aspects of this book is to review the variability in our understanding of *what* and *where* green belts are. Commencing with a brief discussion of the socio-historical influence attached to the poem by William Blake known as "Jerusalem", we discuss this pluralism of our understandings of green belts and explore how personal, communal and even governmental perceptions differ (Cloke, 2006).

The aim of the chapter is to illustrate how, as both planners and members of society, our perceptions of the green belt in the UK could be considered to be outdated. By analysing how nostalgia and memory influence our understanding of the landscape, we can reflect on how we attribute value, which can then be compared to the more mechanistic or procedural ways in which we ascertain quality in the landscape. We then discuss the myriad approaches taken to "valuing" landscapes of "high quality" and ask whether we should continue to promote a uniform approach to the designation and management of green belts.

Green belt size, location and landscape characteristics

All of England's green belts are, by definition, located in close proximity to an urban area, usually the larger ones, although there are some exceptions – for example, Glossop, a small town in south Yorkshire. As discussed in Chapter 2, green belts were created to limit sprawl and prevent coalescence between towns and cities following the Second World War (see Figure 4.1). While the retention of separate and distinguishable urban areas was a key objective of their original designation, green belts were also ascribed the role of maintaining open space around urban areas. Although they were not originally designed to protect specific landscape features or characteristics, since 1955 their ongoing protection in government policy has seen the focus of preservation become more restrictive as the presumptions of what "green belts" are have become increasingly embedded in planning discussions (Amati and Taylor, 2010; Morrison, 2010; Thomas and Littlewood, 2010). Contemporary discussions of green belt designations therefore appear to be an amalgam of historical conceptions of the British – specifically the English – landscape and more nuanced explorations of the potential value that these locations can hold for housing, commercial/industrial, recreational and/or other infrastructure development.

These debates highlight a key dilemma in how we view, research and manage green belt, namely one of understanding. This has established an ongoing discussion in the UK relating to positive conservationist or negative perceived restrictions of the value of the green belt – an issue to which we will return. This implies an understanding of where and what green belts actually consist of. From a reading of literature published by bodies such as the CPRE, 'the green belt' is presented as a uniform set of landscapes which can be categorised as a collective whole (Woods, 2011). This view is pervasive among green belt supporters but fails to take into account the variability of landscapes in and around England's cities. In comparison, the discussion presented by England's Community Forests (2004) and Countryside Agency and Groundwork (2005) regarding England's urban-fringe areas offers a set of ideas, problems and management solutions for these areas which better reflects this variability than do most writings on green belts. As a consequence, therefore, we feel there is a need to identify, discuss and publicise the variability in the location, size and function of the green belt in England if we are to achieve a more balanced view into these discussions. Figure 4.1 outlines the location of England's green belts. It illustrates a clustering of designations around major urban areas and between key urban centres, such as the Liverpool–Manchester corridor. When the post-industrial North differs so much in social, cultural and ecological terms from, for example, the South West, with its small towns, the question arises as to whether we can define green belts as a uniform approach to landscape management. This is an issue to which we will return a number of times in the rest of the book.

Green belt landscapes are characterised as the transitional zones between urban areas, mainly large urban centres, and many rural regions in England. While some designations form visible rings enclosing urban areas, others are spatially more diverse and extend further into predominantly rural areas. Variability is seen in locations where, for instance, the green belt extends extensively into urban locations, such as in Walsall, near Birmingham (Amati and Taylor, 2010; CPRE and Natural England, 2010). Additional complexity is evident in discussions of the location, size and function of the green belt, as a series of towns and cities with large populations are not currently protected by the designation. These locations include such substantial urban areas as Hull, Leicester, Reading and Peterborough. The location of green belts, therefore, does not appear to be a truly robust mechanism for identifying areas of potential growth and coalescence. Furthermore, the dynamism evident in landscape character across the UK undermines the protection of uniform landscapes designations such as the green belt. This raises one of the central questions in current debates: are green belts a coalition of ideas drawn from socio-cultural interpretations or are they characterised by predictability in terms of their form and function? Recent discussions of Areas of Outstanding Natural Beauty (AONBs) have been subject to a comparable

Figure 4.1 England's green belt designations

examination, as the notion of all "special landscapes" is, in some places, being undermined by ongoing discussion related to LPA housing/development objectives (National Trust, 2015). How the landscape is managed, for whom and for what purposes, could be thought of as being a dynamic process that is influenced by a broad range of socio-economic and ecological factors. How we manage such designations therefore provides insights into what priorities or agendas are being promoted within a given development/management context.

There is a wide-ranging literature that examines the nature of these transitional zones, most of which has been developed since green belt policy was first legislated (see Chapter 2). The discussions, however, were predominantly about the value of such spaces rather than their composition or functionality. For example, the more recent publication of the Countryside Agency and Groundwork (2005), entitled *The Countryside in and around Towns* (CIAT), attempted to focus debates on a repurposing of transitional urban-fringe spaces considered to have little socio-economic or ecological value. Reviewing CIAT ten years later, we can identify that the ten benefits proposed for the urban fringe sit somewhat removed from several interpretations of green belts. The most significant difference is the view that urban-fringe areas can, and should, have a functional quotidian purpose, and that people should benefit directly from interactions with these sites. Therefore, although in many areas the green belt and the urban fringe occupy the same location, there is a clear divergence of purpose between utility and protection. Such diversity in the framing of discussions concerning the green belt and the urban fringe helps us to reflect on contemporary interpretations of the locations, as well as the functionality, of green belt resources in England. However, before we can start to assess notions of value and quality, we must first understand where existing green belts are located.

Where are the UK's green belts?

A common myth is that green belts are located around all towns and cities in the UK. However, such an arbitrary understanding of their spatial distribution illustrates one of the problems with how people attribute value to such diverse spaces (Gallent *et al.*, 2006). Fourteen separate green belts are currently designated in England, the largest and most recognised of which is the London (Metropolitan) green belt, covering 486,000 hectares. In contrast, the smallest green belt in England is located around Burton upon Trent and Swadlincote in Derbyshire and covers only 700 hectares. Overall, 13 per cent (1.6 million hectares) of England is classified as green belt, which is a figure many commentators may find surprising, as they consider it to be either too large and restrictive or too small and open to threat (Quilty-Harper *et al.*, 2012). Table 4.1 gives the location of each of England's green belts. Along with Figure 4.1, it highlights how these do not simply encircle urban areas but, as noted previously, extend into more rural areas – for instance, around the Gloucester and Cheltenham area. The physical distribution of green belts in England is not uniform, as large swaths of land in the South West, East, Yorkshire and the North West have no designations. The clustering of green belts around London and its feeder/commuter cities, around Birmingham and its environs, and in the Liverpool–Leeds M62 corridor thus present the largest agglomerations. While there are further designated areas around Bristol, Bournemouth, and Tyne and Wear, it is unequivocally the case that the existing green belts are not located around all cities and do not protect potentially high-quality landscapes in all areas of England.

There are further green belts in the three devolved nations of the UK (see Chapter 2 for more details). In Wales the only formal green belt approved at the time of writing lies between Newport and Cardiff, but there are eleven in Scotland (Scottish Government, 2014) and thirty in Northern Ireland.

The Scottish government, like many commentators (cf. Gallent *et al.*, 2015) promotes a more multidimensional understanding of what the green belt should achieve, similar to Cloke's (2006) discussions of rural pluralism. Its policy notes the supportive and regenerative aspects that green belts can deliver as a

Table 4.1 The location of green belts in England

Green belt area	Area (ha)	Urban areas with 200,000+ residents with green belt	Population	Urban areas with 200,000+ residents and no green belt	Population
Avon	66,868	London	7,215,900	Leicester	303,580
Burton upon Trent and Swadlincote	714	Birmingham	970,900	Kingston Upon Hull	301,420
Cambridge	26,340	Liverpool	469,000	Plymouth	243,800
Gloucester and Cheltenham	6,694	Leeds	443,250	Southampton	234,250
London (Met)	484,173	Sheffield	439,870	Reading	232,660
North West	247,708	Bristol	420,560		
Nottingham and Derby	60,189	Manchester	394,270		
Oxford	33,728	Coventry	303,480		
S-W Hampshire and S-E Dorset	78,983	Bradford	293,720		
South Yorkshire and West Yorkshire	248,241	Stoke-on-Trent	259,250		
Stoke-on-Trent	43,836	Wolverhampton	251,450		
Tyne and Wear	71,854	Nottingham	249,850		
West Midlands	224,954	Derby	229,400		
York	25,553				

Source: Adapted from CPRE and Natural England (2010, p. 15).

key mechanism to support planned or structured economic growth and, like the CIAT agenda, indicates that, through the designation and promotion of green belt resources, town and country can be aligned on a continuum rather than being seen as opposites. The approach taken by the Scottish government appears to be merging the lines between the definitions of urban and rural presented by CPRE and Natural England (2010) and Woods (2011) in a way that many commentators reflecting on England's green belts do not.

When compared to the size of other landscape designations – for example, National Parks or AONBs – the overall scale of England's green belts seems small. Table 4.2 outlines the scale of Sites of Special Scientific Interest (SSSIs), National Nature Reserves (NNRs) and National Parks in England, Scotland and Wales, as well as AONB designations and National Scenic Areas (NSAs) in Scotland. A reading of Table 4.2 suggests, as may be expected, that national scale designations are cumulatively larger in size (ha^2) than the sum of the green belt in England. However, one interesting point to note is the extent to which anti-development/ protectionist influence on policy is visible in both sets of designations.

Due to the variation in the pace of development currently being witnessed in the UK, one of the main arguments presented by the CPRE and others in their

Table 4.2 Key landscape designations in England, Scotland and Wales

	England and Wales		Scotland	
	Ha	%	Ha	%
Nature conservation designations				
SSSIs	1,129,000	7.4	893,000	11.3
NNRs	86,000	0.6	113,000	1.4
National Parks	1,486,400	9.2	–	–
Landscape conservation designations				
AONBs	2,122,510	14.0	–	–
National Parks	1,486,400	9.2	–	–
NSAs	–	–	1,001,800	12.7

Source: Scottish Office (1996).

ongoing support for the green belt is the rapid loss of landscape resources to development (CPRE, 2016). There is an element of truth in this: as discussed in Chapter 2, although between 1979 and 1997 the area of green belt in the UK increased from 721,500 to 1,649,640 hectares (Smith, 2015), there has been a slight fall in recent years (DCLG, 2015). Much of this fall, however, is attributed to the reclassification of part of the green belt in the New Forest National Park in 2005. All of this suggests that the spatial extent of the green belt should not be considered to be static. It also supports our contestation that there is a pluralism to green belt discussions which reflects a misunderstanding of their size and location.

A further question raised in discussions about the green belt reflects on the co-location of their *value* and our understanding of what constitutes a *high-quality* landscape designation. The notion of value is a key aspect of the CPRE's rhetoric for the protection of the green belt (CPRE and Natural England, 2010). However, we can, and should, question whether the green belt as a whole in the UK is of high quality. This includes understanding where green belts are intersected by National Park, SSSI, Country Park or other landscape designations. Currently only a small area of the green belt is designated as either a National Park (just 84 hectares) or an AONB (9 per cent of the total). Moreover, the amount of land designated as an AONB varies considerably between green belt areas. Some have none (Cambridge, York, Nottingham and Derby, Stoke-on-Trent and Burton upon Trent and Swadlincote), while a quarter of the London Metropolitan green belt around London and more than 20 per cent of the Avon green belt is classified as AONB (CPRE and Natural England, 2010). Furthermore, 33 per cent of Local Nature Reserves in Great Britain and 44 per cent of England's Country Parks are located within the green belt, suggesting that, to some extent, these landscapes can be categorised as being of high quality. However, other stretches of the green belt, such as areas in the Merseyside designation, could be classed as being of a lower quality and lacking in ecological or cultural significance. Our understanding of "quality" as a set of quantifiable classifications thus appears to influence the rhetoric of such discussions in green belt areas.

Establishing green belt *value* and *quality*

Extending such arguments of *quality* and *value*, one of the key issues with the current role of green belts is how we define and subsequently plan against "inappropriate" development. While each individual will have a different view of what constitutes "inappropriate", the UK government has become more pre-scriptive with its own interpretation (DCLG, 2012). The House of Commons South East Regional Committee expressed the following view:

> There are advantages to the green belt policy and it is undeniable that it has helped to retain the rural character of large areas of the region which other-wise would have become overwhelmed by urban sprawl in the last 50 years. However, it was designed for a different time, and it is now working against the ideal of sustainable communities which hope to encourage people to work, rest and play in the same local area. As a result, there are areas of the region where the green belt is adding stress to the immediate transport net-work and inadvertently placing pressure for development on valuable areas of greenspace within urban areas. We recommend continuing support for this policy of selective review of green belt in the South East Plan.
>
> (2010, p. 21)

In contrast, the CPRE regularly reiterates that it feels that green belt policy remains a highly effective form of managing development and protecting the environment in the countryside (CPRE, 2015a). The organisation has argued that it would like to see an extension of green belt policy and the physical pro-portion of the landscape classified as green belt to encapsulate additional areas, to ensure that the countryside of England retains its rural aesthetic (CPRE and Natural England, 2010). The CPRE's view contrasts with that of authors such as Paul Cheshire (2013), who argues for a more pragmatic approach to green belt discussions. Cheshire proposes that, as a country, the UK should attempt to limit the nostalgic lens through which we view designations and instead build on areas with limited ecological and social functionality. He also suggests that, if the sub-jective nature of value can be accounted for, planning regulations should be used to protect more ecologically and socially valuable areas. He bases this discussion on the view that we need to take a more rational approach to protection which identifies specific values and functions for retention rather than continuing to utilise a blanket reluctance to allow any development. Furthermore, he notes that some of the "scrubland" currently classified as green belt – for instance, in Tyne and Wear – could be released, as the landscape is not necessarily of high quality or great biodiversity and provides only a small number of homogeneous benefits.

We can therefore debate whether the relationship between green belts and spa-tial location in the UK should be discussed in terms of *quality* or *value*. Although urban-fringe transitional zones are important stepping-stone landscapes between urban and rural areas, we propose that a homogeneous value should not be placed on these spaces. Alternatively, the location, the physical characteristics and the nature of agricultural and commercial use of the green belt suggest they can be viewed as being heterogeneous spaces. The hybridity of activities and perceptions of rural landscapes, and by extension the green belt, presents a complex set of processes which influence value (Cloke and Little, 1997). For instance, the green belt running east from Liverpool includes what could be described as typical English "chocolate box" villages (see Figure 4.2) but also areas which are characterised by lower standards of environmental quality (Pauleit *et al.*, 2005). Both, though, are part of the same "valuable" green belt in Merseyside.

Furthermore, if we review some of the terminology used by green belt "protectionists", we could consider the green belt to be fragile, vulnerable and disordered – labels that do not describe all the green belt in Merseyside – and therefore in need of protection; a large percentage of the Merseyside green belt is productive agricultural land, which is ordered and managed accordingly. Extending such debates, we could propose that the homogenisation of the green belt within popular discussions is drawn from a legacy of landscape valuation, which does not necessarily represent more contemporary understandings of locality and people or the perception of urban–rural interface landscapes (Halfacree, 2006). This is problematic in the UK, as rural landscapes are imbued with historical value, many of them reflecting a romantic and nostalgic appreciation grounded in heritage studies or literature (Matless, 1998). These views

Figure 4.2 A house in the village of Hale in the Merseyside green belt

are prominent in the language used by campaigning groups, the UK government and others (Sturzaker, 2010). We suggest that such a homogenised approach towards the discussion of green belts limits their utility and "value" to the UK in the twenty-first century.

If we take this argument further and search for a more pluralistic view, then we can find those who suggest that green belts are a continuum of spaces that range from high-quality National Parks to low-grade urban-fringe scrubland (Davies *et al.*, 2006; Mell, 2010). As a consequence, it might be more rational and practical to manage these landscapes as individual sites rather than as a collective entity. Currently the UK government, through the NPPF, sets out a uniform approach to the management of green belts, offering little flexibility for LPAs to meet local needs – for example, for housing in rural areas (Satsangi and Dunmore, 2003). However, the variability of green belt locations does not really allow for such a uniform approach to management because it diminishes the ability of LPAs to react to contextual differences in local planning objectives and environmental quality (Gallent *et al.*, 2015). On a more global scale – a view to which we will return in Chapter 5 – we can identify similar issues in international green belt discussions. For instance, the cities of New Delhi and Ahmedabad in India both have green belt areas which were set up to moderate the process of urban expansion. However, because of the rate of growth and the variability in local land use, the politics of landscape planning, and land-scape composition, areas of the green belt *have* in fact been deleted to meet local needs (see Chapter 2). Politicians, planners and developers in Ahmedabad and New Delhi in India have struggled to balance these issues (Mell, 2016). Although this process may be socio-economically led, it does highlight that flexibility can be integrated into green belt discussions, although such flexibil-ity is normally centred on economic (or social) uses and less so on the protection of the natural environment. In the green belt discussion centred on Milan (Italy), for instance, there appears to be a more nuanced appreciation of landscape quality. This illustrates the capacity of the green belt (and its extensions into urban forestry) to increase the ecological functionality of the city's environ-ment through a network perspective (Spanò *et al.*, 2015). The green belt is also designed to act as a mechanism to increase connectivity between the city of Milan and nearby towns. Milan's use of the green belt therefore suggests that a more holistic ecological imperative can be aligned with the promotion of socio-cultural values. To a lesser extent, the use of the green belt in India highlights one of the CPRE's main concerns for these resources – that alternative and, in many cases, economic imperatives can start to override environmental consid-erations as development pressures increase. In a UK context, the possibility of opening up green belt land for development is often seen as anathema to the visible protectionist agendas of some campaign groups (see Figure 4.3). We propose, though, that a more locally contextualised approach to protection and/ or release could lead to a more rational and appropriate form of management for green belt areas.

Figure 4.3 Opposition to green belt deletion in Merseyside

What is landscape value?

Value is an intensely personal matter. Valuing a landscape or a landscape feature is therefore variable, dynamic and evolutionary in how we assess a resource, as well as being jointly static in terms of what we value (Tyrväinen and Väänänen, 1998; UNEP–WCMC, 2011; South Yorkshire Forest Partnership and Sheffield City Council, 2012; Mell *et al.*, 2013). As noted above, attempts to debate development in the green belt are characterised in the UK by opposition. Central to this is an ongoing proposal that green belt areas are of *high quality* and, by association, of *high value*. However, value is an extremely subjective notion – especially when landscape values are being explored. Value is a pluralistic concept based on the integration and interaction of experience and knowledge of a location, its physical structure, and its function within a given geographical area. Therefore there is no "correct" way to value a landscape resource. Valuation is, rather, a constantly evolving set of ideas, principles and socio-cultural interpretations of what makes a location important (Tuan, 1990; Nassauer, 1995). Consequently, there is no singular or overarching view of what makes the green belt valuable based solely on its ecological, social or economic composition. This is not necessarily a problem, as it provides greater scope for planners or landscape specialists to allocate value to areas for different reasons and to support alternative meanings

(Herrington, 2009). How we view a given landscape, what we see in it, and how this has affected our knowledge and experience of that location could thus be more important to valuing a landscape than its designation.

It is also possible to extend this idea to consider green belt as a hybrid concept. Green belts are clearly spatially defined, being marked on maps and plans. We could, however, debate whether their value(s) is (are) as clearly defined. Such a pluralistic approach to green belt discussions, though, has proved inconvenient for many commentators (Amati and Taylor, 2010; CPRE and Natural England, 2010; Evans and Freestone, 2010), as it implies that there is a variability to what can, and should, be classified as a green belt. Therefore, as we noted previously, our knowledge of where and how large green belts are, and what value they have, could be viewed as being socially constructed, and potentially as socially skewed (Tuan, 1990). Unfortunately, this places green belts in a difficult position legislatively and adminstratively. To protect urban-fringe or rural landscapes in the UK, the government created specific green belt policy with the aim of limiting the level of growth in these areas. However, unlike the case with other environmental designations, there is no well-defined set of characteristics setting out what the green belts are. Furthermore, planning policy could be viewed as providing an underlying ethos for such discussions based on our cultural understandings of the landscape rather than for a specific location (Woods, 2011). Subsequently,within such discussions, value can become a byword for environmental protection and a retention of what we historically or nostalgically think of as the green belt. On the other hand, when we look at designations relevant to the planning system relating to biodiversity, such as Special Areas of Conservation or Ramsar sites, we are presented with a clear set of constitutive principles (Cullingworth *et al.*, 2015).

We also propose that the pluralistic nature of value could be viewed both qualitatively and quantitatively. In much pro-green belt literature there is a clear rhetoric surrounding their innate or experiential value which is difficult to argue against, because the metrics used to discuss "experience" are relatively vague (Roe and Taylor, 2014). This is in contract to the more prescribed mechanisms used by Natural England and other environmental organisations to designate areas of ecological or socio-cultural value, such as SSSIs. Such practices are far more prosaic in terms of what geological, topographical or climatic factors contribute to value. They are also more detailed in what kinds of socio-cultural activities support a designation of a valuable landscape. Some designations – for instance, Local Character Assessments (LCAs) – integrate experience into these discussions and provide a much more detailed approach to valuation (Natural England, 2010). We can therefore ask whether we should be focusing more directly on the establishment of criteria to attribute value to green belt landscapes or whether we should continue to work within the pluralistic approach we currently utilise. If we identify this dichotomy of qualitative vs. quantitative practice within our valuation, the variation illustrates why green belts can be considered as being contrived, contested and convoluted, as well as a context-specific planning issue. Furthermore, returning to the research of Cloke (2006), we can argue that the understanding of value across the urban–rural interface is fraught

Table 4.3 Statutory agencies and their landscape management responsibilities

	England	Scotland	Wales	Northern Ireland
Statutory advisor	Natural England	Scottish Natural Heritage	Natural Resources Wales	Department of the Environment Northern Ireland
Principal landscape designation	National Parks	National Parks	National Parks	National Parks are under construction
Specific landscape designation	AONBs, Heritage Coasts	National Scenic Areas	AONBs, Heritage Coasts	AONBs

with conjecture and subjectivity, which reflects the UK's broader discussion of landscape functionality. All of which makes it difficult for planners and other stakeholders to align culturally constructed interpretations of value with more pragmatic attempts to quantify landscape quality.

Table 4.3 shows how different locations integrate added complexity into the process of quality assessment. If we consider which organisations are tasked with acting as statutory advisors to LPAs for landscape protection, we can identify a diverse series of actors. Each of these stakeholders – for instance, Natural England – utilises its institutional agenda to frame its discussions of landscape value. However, the administration of such responsibilities is complicated, as landscape features, including green belts, often extend across administrative boundaries. Ensuring that evidence-based decision-making is undertaken through collaboration is one of the key drivers of the ongoing engagement of environmental considerations in statutory planning. Moreover, attempts to comply with the European Landscape Convention's promotion of value in quotidian landscapes, and not simply in rare or protected ones, makes management a more convoluted process (Roe *et al.*, 2009). How we manage the range of landscape designations within planning policies and delivery mechanisms therefore makes it difficult for planners to promote an overarching appreciation of what value should be given to green belts. It is within such debates that campaign groups utilise policy vagueness to establish their platform for protecting "valuable" landscapes without necessarily defining what is valuable about them.

What is landscape quality?

Landscape quality can be conveyed through social, economic and ecological discussions. In some locations and disciplines – for instance, landscape ecology (Jongman and Pungetti, 2004) – value is associated with the connectivity and support of ecological networks, while in rural studies and cultural geography we see more socio-economic and cultural factors being discussed (Cloke and Little, 1997; Woods, 2011). Furthermore, within the planning and environmental management literature the economics of landscape is proposed as being a pertinent factor in understanding value (Gallent *et al.*, 2015). However, value and quality, as discussed previously, are different, and this difference affects how we perceive, utilise and manage the landscape. Discussions of green belt form a prominent example of how the differences between quality and value can become blurred depending on the context of the debate.

For example, the Scottish government (2014) places value on the ecological elements of the landscape in its green belt policy. It also emphasises the added value that green belts and sustainable development make to the Scottish economy and its cultural heritage and goes further, stating that all public bodies, including LPAs, have a duty to enact and promote the conservation of biodiversity under the Nature Conservation (Scotland) Act 2004. This needs to be integrated into development plans and decision-making at all levels. Furthermore, biodiversity is presented as an important element of green belt designations in Scotland, as it provides both ecological and social services and makes a significant contribution to the Scottish economy. Conversely, in England, the NPPF states that the 'fundamental aim of green belt policy is to prevent urban sprawl by keeping land permanently open; the essential characteristics of green belts are their openness and their permanence' (DCLG, 2012, p. 19). Green belts in England are therefore subject to a far more restrictive definition of value in terms of the prevention of coalescence. However, this is then centred on the need to protect the openness and aesthetic quality of the landscape – a socially constructed notion – as well as noting the permanence of green belts. This illustrates some of the similarities to the proposed values presented by the Scottish government in terms of social and cultural heritage being important. However, the NPPF lacks the same explicit ecological perspective. Moreover, the 'five purposes' of green belts outlined in the NPPF (see Table 2.2) suggest that socio-cultural interpretations are important.

The contrasting approach to green belt discussions at the national level in England and Scotland illustrates the complexity of establishing what green belts are, what they should do, and how we value them (Amati and Taylor, 2010). In the UK, however, there are a series of evaluative mechanisms which have been used to identify quality in the nation's landscape.

One approach to identifying quality proposed by Natural England and by the UK government within the NPPF has been the use of Landscape Character Assessments (LCAs) and, latterly, National Character Areas (NCAs) (Natural England, 2010; DCLG, 2012). The rationale supporting LCAs and NCAs sought to describe the character of the landscape and identify the unique combination of elements or features that make an area distinctive. Character is considered to include topographic, ecological and socio-economic variables, as well as experiential values associated with culture, history and memory (Natural England, 2010). A significant aspect of LCA and NCA designations was that, unlike the perceived static nature of green belts, they evidenced the changing nature of the landscape. By examining how this was happening, what socio-economic activities were supported, and what environmental opportunities could be supported in the future, these designations highlight the dynamic nature of establishing notions of quality in the UK's environment. The relevant discussions were based on landscape features and were developed by following a "systems" approach to designations rather than a technical LPA/administrative boundary approach. It looked at the wider value of an environmental system to form landscape boundaries rather than, as with green belts, an avoidance of coalescence. LCAs have subsequently been scaled up by Natural England to cover the entire country. The 159 individual designations are discussed in the NPPF as a process that helps shape planning to recognise the intrinsic character and beauty of the landscape within local plans (DCLG, 2012). However, like other landscape designations in England, and in the wider UK, NCAs and LCAs could be considered as having a partial protective value because they do not hold the same level of legal responsibility as National Parks or AONBs.

In 2010 the CPRE and Natural England released a report assessing the current quality of landscape in England's green belt. They made use of the Countryside Quality Counts (CQC) project to evaluate the state of England's green belts and shifted the emphasis placed on them from the NCA to more specific LCA approaches. Using the NCA parameters, the CQC proposed four character types that were felt to be consistent with quality: two positive – maintained and enhancing – and two negative – neglected and diverging (see Table 4.4 and Figure 4.4). From the analysis, 39 per cent of green belt land was considered to be in a stable/maintained state and with an established landscape character – i.e., a character perceived to hold a specific quality. In addition, 36 per cent of the green belt land assessed could be classified as holding an emerging landscape character, a significant proportion of which was associated with post-industrial change/landscapes around the core cities in the north of England. One caveat to this analysis is that the NCAs do not match the green belt boundaries. This meant that the analysis had to take a proportional perspective whereby the percentage of the NCA lying within the green belt was evaluated against the four categories noted above.

The four classifications created by the CPRE and Natural England help to frame discussions of NCAs/LCAs within an evolving landscape context. Although change can be considered as differing in a negative sense from the situation in the late 1990s, the suggestion is that what was considered unique or meaningful at that time may have been eroded or evolved further, while the positive characteristics are framed as extending the "value" of a given location.

Table 4.4 National Character Area (NCA) and change

	Consistent with character in the late 1990s	Inconsistent with character in the late 1990s
Stable	*Maintained*	*Neglected*
	Character is strong and intact. Changes observed serve to sustain it. Lack of change means qualities are likely to be retained.	Character of area weakened or eroded by past change, or changes observed were not sufficient to restore qualities that made the area distinct.
Changing	*Enhancing*	*Diverging*
	Character has restored or strengthened character of area.	Change is transforming character so that distinctive qualities are being lost or new patterns are emerging.

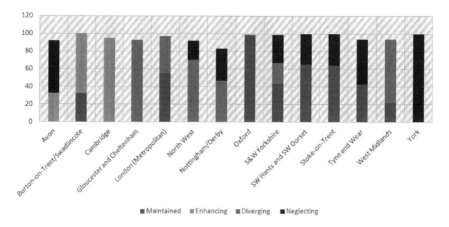

Figure 4.4 National Character Assessments of the green belt in England

The terminology used in such discussions can, though, be mapped onto corresponding debates surrounding green belts.

The use of unique and intrinsic values again places a socio-cultural value on the landscape, be it at the level of an NCA or at a smaller scale – for example, the green belt around a particular town. In addition to issues of scale, there is a need to discuss the character of the landscape and how these features influence perceptions of value. Natural England's guidance (2010) suggests that there appears to be a clearer set of parameters which help designate NCAs as compared to green belts. This is based on the development of a distinctive or unique set of landscape characteristics – hence the 159 designations. In contrast, green belts have the same level of protection wherever they are, despite the clear variability in the quality of land. Moreover, if we assess the planning restrictions placed upon green belts compared to National Parks or SSSIs, we could argue that there is a high level of complementarity between them. Such disparities between the scale and function of different designations does not, however, appear to be have been debated to the same extent as their protection. Furthermore, even within a more structured evaluation, the CPRE and Natural England found it difficult to draw conclusions from their analysis on account of the variability of the landscapes. It may therefore be more practical to use a number of alternative methods and techniques to identify what we mean by "quality" in the green belt, and there are questions as to whether we can value and manage green belt areas as a collective entity rather than locally contextualised landscapes framed by social and economic interactions or interpretations of the environment.

Green belts: valuable or high-quality landscapes?

As discussed above, the ways in which we value *anything*, including the green belt, are very subjective, which leads people to interpret landscapes in very different ways (Herrington, 2009; Roe and Taylor, 2014). In part this reflects the ways we attribute "value" and is dependent on a myriad of alternative socio-economic and ecological variables, many of which are not compatible with each other (see Chapter 6 for a further discussion). *What* we value also has a major influence on *how* we manage the landscape. We therefore need to reiterate that the value of green belt is not the same as its quality, as many areas of green belt could be considered to be low quality (Papworth, 2015). The differences between value and quality should therefore be at the centre of green belt debates. Unfortunately, our understanding of these variations is often subsumed within wider arguments about what we can and cannot develop in green belts. If, however, we view value as being a socially constructed mechanism designed to appropriate characteristics we find beneficial, this can be counterposed with examinations of quality which have a set of measurable parameters. This raises the differences in protection offered to landscape designations, illustrating the variability in how these issues are debated and/or forcibly enacted. Moreover, there is a secondary element to such discussions which argues that, if the attribution of value is a socially constructed process whereby we place implicit value on certain landscapes because we consider them to be historically or culturally valuable, we are also saying that other landscapes lack value. This runs counter to the commentary of the European Landscape Convention and potentially also limits how we discuss the quality of other quotidian landscapes which are not classified as green belt. Within the literature on green belts there is little discussion of how we balance these competing issues (Roe *et al.*, 2009).

To explore whether value and quality are comparable in green belt discussions, we can look at the locations of these resources compared to those of other landscape designations, such as National Parks or NCAs. The latter designations are related to, and help identify, qualities in the landscape but are often associated with innate values, rarity or scenes of historic beauty (Nassauer, 1995; Matless, 1998; CPRE and Natural England, 2010). What they do not consider is how we differentiate between value and quality. Furthermore, if we take the opposing view of green belts as areas of low-quality land that supports only minor ecological or socio-economic value, can we describe them as being of high quality? Within the broader rhetoric of green belt discussions, these landscapes are still seen as valuable as they are areas which require protection from development because they are classified as green belt. It would appear that the issue of quality does not always enter into this conversation. Similarly, when we consider other landscape designations (for example, SSSIs), we become increasingly aware of the subject of quality when adapting legislation to protect resources. Partly this is to meet the legal checklist of requirements that govern what we should and should not classify as high quality within local, UK-wide and international policy (Cullingworth *et al.*, 2015). However, designations such as National Parks in the UK are not constituted of 100 per cent high-quality features but are a mosaic patchwork of landscapes that are collectively deemed as valuable. Many of the areas at the periphery of the Lake District National Park, for instance, could be considered to be of relatively low quality. However, when development is discussed in this area, the cumulative impact on the wider value of the area is taken into consideration (Matless, 1998). This is not replicated in green belt discussions.

Given the variation in how the landscapes of the UK are valued, we could ask whether this represents a continuation of the normative historical understanding of the environment. A number of authors have discussed how the landscapes have been shaped by our valuation of the arts – paintings, poems and literature (Lowenthal, 1985; Brace, 1999; Gallent *et al.*, 2004; Herrington, 2009). Such discussions frame the landscape as a valuable entity in totality but do not always reflect the broader deliberations about the changing needs of development. While we could argue that development in a National Park would need to be sympathetic to the landscape aesthetic, there are questions of social justice relating to the stance that new homes cannot be built to maintain communities because they may diminish a high-quality landscape (Richards and Satsangi, 2004). Scaling this argument to the green belt, we might ask whether it is, in total, all of high quality, or whether there are areas of high quality that are valuable ecologically, socially and economically as well as areas of low-grade agricultural land (Papworth, 2015). For example, the visual amenity value of the green belt in and around cities such as Liverpool could be considered, in places, to be of low quality (see Figure 4.5). However, even in such areas the reactions to green belt release is negative, as it is deemed to be removing a valuable local resource (see Figure 4.3, which is in close proximity to Figure 4.5). As a consequence, we see a repetition of the protectionist rhetoric which has been pervasive in the UK over a number of years.

Figure 4.5 Some lower quality green belt in Merseyside

If we were to review the quality of the green belt using Landscape Character Assessments or against such landscape designation criteria as Special Protection Areas or SSSIs, we might see that a large proportion of the green belt is not of high ecological quality. What we must remember, though, is the view that many communities identify green and open spaces as valuable because they believe such spaces support a better quality of life (Roe and Mell, 2013; Clark, 2015). This appears to be the case even in locations where the physical composition of the landscape may look low grade to non-residents. It may therefore be a useful exercise to assess green belts through a more quantitative lens to establish what *landscape quality* is actually present and what could be considered to be socially constructed.

One method of identifying the quality of the landscape is to use Landscape Character Assessment. Landscape character can be defined as a distinct and/or recognisable pattern of elements, or characteristics, that allows the observer to make distinctions between locations. This is a test of distinctiveness rather than describing which landscape is better or worse (Tudor, 2014). LCA is thus a process by which we are able to identify what is *good*, *bad* and *different* within the natural environment. When conducting an assessment, a checklist of features (natural, cultural and socio-economic) is used to frame what characteristics are deemed to be distinctive. Although the assessments are more mechanistic compared with some other forms of landscape valuation, there is still a clear mandate within the process to include more subjective interpretations of the environment.

Given the nature of green belts, the use of an LCA could provide a supportive framework both for defenders of these areas and for pro-development lobbies. Using such an assessment may offer a set of foundations that people believe they understand or can buy into in terms of the location, function and amenity value of the green belt. Included is the incorporation of an understanding of landscape heritage and diversity and how such narratives can be integrated into modern decision-making (Countryside Commission, 1994). Landscape Character Assessment can thus be used to inform our knowledge of what the landscape is comprised of and help planners to make judgements on its value. When conducting assessments it is important to reflect on the following five areas.

1 Landscape is everywhere and all landscape and seascape has character.
2 Landscape occurs at all scales and the process of Landscape Character Assessment can be undertaken at any scale.
3 The process of Landscape Character Assessment should involve an understanding of how the landscape is perceived and experienced by people.
4 A Landscape Character Assessment can provide a landscape evidence base to inform a range of decisions and applications.
5 A Landscape Character Assessment can provide an integrating spatial framework – a multitude of variables come together to give us our distinctive landscapes.

(Tudor, 2014, p. 12)

Using mechanisms such as LCAs or NCAs to identify distinctive and functional locations provides scope to add a quantitative aspect to how we view the landscape. It also helps us to question whether green belts are uniformly valuable and of high quality. Although a number of green belt locations have a distinctive character because of their location near to or within other landscape designations, this does not necessarily mean they are of high quality. The examples given from Merseyside suggest that it may be short-sighted to limit development based on an assumption of "landscape value or rarity" in the green belt, especially where landscapes could be considered to be of low grade. We might therefore ask what the perceived distinction might be between those areas we deem fit for development and those which require protection in perpetuity? A range of authors have approached this issue from a moral, an economic and a landscape perspective, questioning whether the ways in which we perceive the landscape is limiting our ability to manage it effectively (Cloke, 2006; Halfacree, 2006; CPRE and Natural England, 2010; Woods, 2011; Tudor, 2014). The following sections extend some of these arguments, drawing on the more experiential literature to examine how experience, memory and nostalgia might play a role in how we frame our discussions of green belts.

Contextualising perceptions of green belts

The future of green belt areas needs to be discussed in the wider context of land use. How people understand the pluralistic nature and socio-economic value of green belts impacts directly on their position they take in such debates. From the British angle, we can identify a lineage between historical perceptions of landscape value and current policy (Healey and Shaw, 1994). However, to unravel the connections between people, the landscapes of the UK and green belts can be problematic, especially in locations deemed to have important landscape character. The processes of valuation have been shaped in various ways by allegorical, socio-cultural and, in some cases, elitist views, encapsulated in the current interpretations in the UK of William Blake's famous poem (*c.* 1808):

> And did those feet in ancient time
> Walk upon England's mountains green?
> And was the holy Lamb of God
> On England's pleasant pastures seen?

So starts one of Britain's best-loved poems – verses that evoke very British feelings – which places allegorical subtext at the centre of our relationships with the landscapes around us. Blake's verses, in the form of the hymn into which they were adapted, are played at weddings and sporting events and are verses which some consider illustrate the need to conserve *ad infinitum* the remaining "pristine" landscapes of the UK. However, it might be argued that such an understanding of the landscape, and by extension the green belt, is somewhat illusory. We can therefore with alacrity examine how perceptions of green belts are constructed by the ways in which, in the UK at least, landscapes have been described, painted and subsequently valorised. This raises the question posed by Woods (2011) from a nostalgic point of view as to *when* is the rural or green belt rather than *what* these resources are?

The perceptions of Britain's landscape that have been borne out over years of discussion within academia (Matless, 1998), village halls (Morrison, 2010) and campaign literature (Sturzaker, 2010) are as nuanced as they are diverse. The classic view that the British countryside, and, in the context of this book, green belt, is one endless summer of afternoon teas, cycling on country lanes and cricket on the village green was perhaps never accurate and is certainly now heavily contested (Matless, 1998). Denis Cosgrove's 'landscape idea' explored this notion reflecting on the breadth, and subsequent role, of change in the needs of the public to perceive social, economic and ecological evolution of the landscape (Cosgrove, 1985). Extending this view, Cosgrove proposed that, although we learn from the past and are capable of adapting our practices and discussions to current phenomena, we also recognise that we have an innate ability to retain the value of historical landscape legacies. Our understanding of the countryside, as individuals, and professionally as planners, although couched in these terms, is now required to take on a more hybridised form (Woods, 2011). What we see, and how we view the landscape, is no longer what our parents saw or what our children will see. Blake's verse belongs to a nostalgic era of idealism, which many would argue may never have actually existed (Spirn, 1998), as do the poems of Wordsworth or the paintings of Constable or Gainsborough. Moreover, David Lowenthal, in his seminal review of landscape and social history, quoted L. P. Hartley's famous words 'The past is a foreign country: they do things differently there' (Lowenthal, 1985). In his work Lowenthal questions the rationale behind our desire to hold on to interpretations of the past, asking whether a refusal to embrace modernity resigns us to repeating the actions of our predecessors. What Lowenthal is suggesting is that the notions of landscape preservation are potentially flawed. In terms of green belts, we can reflect on the time of their designation and the differences between the UK in the post-Second World War era and that of the twenty-first century. All of which brings us face to face with a dilemma: how can we manage the conservation or development of green belts when there is a very nostalgic elephant in planning committee rooms?

Green belts: imagined or reality

Over the course of the twentieth century, our relationship with the landscape shifted significantly from one of majesty to one of utility (Gilg, 1996; Woods, 2011; Gallent *et al.*, 2015) – a notion which was extended after the Second World War, when the UK reconsidered its productive–consumptive relationship with the landscape (Macnaghten and Urry, 1998). We must reflect on historical changes to landscape functionality if we are to understand where we currently are. One key driver of this change was the UK's movement from a predominantly pastoral nation to one of mechanised agriculture. Throughout the eighteenth and nineteenth centuries the interactions of people with the British countryside became more than a location or a job (Gilg, 1996). The countryside flourished during the Industrial Revolution and allowed the expansion of transport networks and an industrialisation of the land; it also enabled many people to think more reflectively about the aesthetic qualities that landscapes possess. A comparable process took place in North America with the growth of car ownership, the greater amount of free time and the development in the early 1900s of parkways and, later, greenways (Little, 1990). The traditional conception of what constitutes a "countryside location" therefore started to become blurred, as, even during the era where proposals emerged to integrate (or "marry") town and country (Howard, 1898), we started to see a distinctive shift in the long-term composition of rural areas. We also started to conceptualise green belts as intermediate locales between town and country which offered complementary poles of influence for both (Woods, 2011). The green belt could thus be proposed as a mechanism for addressing the fragile, vulnerable and disordered aspects of rural England, as discussed by Woods and the CPRE, in order to move beyond a constructed representation of these resources to a more nuanced or ascertained reality (Mormont, 1990). This issue is one of the main points of discussion in the contemporary literature and campaigns relating to green belts.

From the mid-twentieth century onwards, this process became more evident, as the UK's countryside became increasingly mechanised and populated by traditionally urban communities (Hall and Tewdwr-Jones, 2010; Cullingworth *et al.*, 2015). Migrants to rural areas therefore contrasted with the more traditional flows of people to towns and cities witnessed in the period up to around the Second World War. As our urban areas burgeoned, we saw the start of a counter-urbanisation movement in which people choose to move to rural areas. A further consequence of this process was a growing concern over the retention of the countryside as distinctly different to urban areas, which manifested in the need to avoid coalescence and the development of green belts. We have thus witnessed a blurring of definitions, and, by association, our understanding of town and country, urban and rural, has enabled individuals and groups to appropriate a socially constructed narrative for green belt protection that could be considered to be imagined or nostalgic (Cloke, 2006, p. 18). How we differentiate between the quotidian understandings of green belts compared to the pluralistic notions of Halfacree's (2006) *locality, representation* and *people* model is therefore increasingly difficult. The conceptual representations of the green belt therefore appear to follow Spirn's (1998) proposal that landscapes are a metaphor for social activities and not necessarily representative of their actual value. Any discussion of the green belt should thus contend with the application of a rationalistic view of planning and management alongside the more ethereal concepts of landscape value.

One exponent of the rationalistic role of green belts was Patrick Abercrombie, who envisaged a future 'with a continuous green background of open country in which are embedded at places compact spits of red, representing buildings' (Abercrombie, 1945, p. 20). As discussed in Chapter 2, as part of his development of the London Plan, Abercrombie conceptualised a form of spatial planning where urban and rural would be separated by green rings. Abercrombie could therefore be considered to hold to quite distinctive views on the UK landscape. He felt that urban areas were economically driven and needed to address the requirements of industry, people and commerce, though he also argued that planners needed to retain the "natural" breaks in the landscape to ensure that the pastoral interludes of a bygone era were not lost (Hall, 2002). He therefore saw green belts as an embodiment of the separation of urban and rural. However, Abercrombie could also be viewed as making contemporary references to Howard's garden cities ideal as a mechanism to ensure that people and place retained an orderly position (Matless, 1998). This reflects the ordered assumptions made by Halfacree (2006) in his alignment of people, place and understanding with the rationalisation of rural areas made by Healey and Shaw (1994), who discussed the challenge of economic liberalisation to the more nostalgic interpretations of green belt areas. What was alluded to by both Halfacree and Healey and Shaw was the difficulty that planners, developers and the public have in differentiating between valuable landscapes and working/functional landscapes.

However, it might be argued that the rationale for what constitutes *countryside* and, by extension, what we understand by green belts, bears less resemblance to the imagery of British poets and landscape painters and more to that in the works of contemporary writers or artists such as Ken Russell or L. S. Lowry (Herrington, 2009). We also suggest that the proposals of Abercrombie for London and the initial green belt designations have to some extent marginalised discussions of their utility in the twenty-first century (Amati and Taylor, 2010; and see Chapter 2). We therefore live in an era when hybrid visions of nostalgia appear to reflect a collective need to frame our current relationships with the landscape through a lineage of past engagement. This appears diametrically opposed to the alternative narrative of green belts, which views them as functional landscapes that should be encouraged to diversify to meet socio-economic and ecological needs.

Paul Cloke's exploration of the hybridisation of urban and rural, and, by extension, the green belt, raises a number of interesting contestations, some of which have already been discussed in this chapter (Cloke, 2006). One of the most prominent is the relationship between the imagined and more realistic understandings of how people interact and value the green belt. We can delve deeper into this to examine again the notion that landscapes are socially constructed (Pepper, 1996). As noted above through the commentary of Lowenthal and Cosgrove, what we consider to be a valuable landscape is, potentially, extremely contrived. As a consequence we need to reflect on the factors which help to define what we mean by green belts, how this affects their spatial form, and whether attempts are being made to balance the imagined with the reality. Drawing on the work of Matless (1998), Herrington (2009) and Spirn (1998), we can interpret landscape as a metaphorical lens through which we define order, value and understanding of the world around us. Within the current debates on green belts in the UK, this manifests in the static reflection of pastoral landscapes imbued with historical meanings (Amati and Yokohari, 2006). However, as is inevitable, change occurs, and the urban-fringe and rural landscapes in the UK are in a state of flux. Our reliance on nostalgic narratives to frame green belt discussions lacks an understanding of how socially constructed interpretations can, and should, also change over time (Pepper, 1996). As Healey and Shaw (1994) discuss, the neoliberalisation of the British landscape has led to a shift in emphasis on how we manage the environment and what activities are allowed there. Moreover, returning to Lowenthal and Hartley's description of the past as a foreign country, as well as Matless's discussion of the English landscape, we can see significant changes in how we perceive green belts and how we allocate value to them even if policy has not diversified.

Researchers exploring the value of nature in landscape studies and environmental or behavioural psychology, such as Nassauer (1995), Tuan (1990) and Kaplan and Kaplan (1989), debated these notions and identified coalitions of values that frame our understanding. While these focus on the role of communal constructions of value associated with heritage (CPRE and Natural England, 2010), they also review the role played by the media and the arts in shaping our appreciation (see also Lowenthal, 1985; Schama, 1996; Herrington, 2009). Such a pluralistic debate could, in many ways, dilute the actual discussion of what green belts are for, but because of the seemingly entrenched discussions of their value it has become, as with the wider rural landscapes of the UK, difficult to separate rhetorical views of green belts from the actual needs of the populations who live there. As a consequence we see less exploration of contemporary understandings of the landscape, such as those proposed by Cloke (2006) and Macnaghten and Urry (1998), being engaged with in the wider examination of green belts. This has, unfortunately, led to disagreements within policy-making environments where pro and anti-green belt campaigners have attempted to influence the focus of UK policy.

Imaginary green belts and the National Planning Policy Framework (NPPF)

Nowhere in planning are disagreements about planning policy and practice more evident than in the arguments over green belts, which remain enshrined in the NPPF (DCLG, 2012). Pro-free market advocates, though, are increasingly proclaiming that protected lands are stifling economic growth (Cheshire, 2014; Papworth, 2015), while campaigners, both rural and urban, are likely to protest that *any* development in the green belt is an infringement of their inalienable right to protect our green and pleasant land (see Figure 4.3). While this book does not propose to rekindle, reframe or directly engage with existing discussions on opposition to development (often pejoratively dubbed NIMBYism) or, indeed, rights to the countryside, as there is a vast literature on these issues (Wolsink, 2006; Schively, 2007), we aim to extend the examination of the validity of the two main voices visible in green belt arguments in the UK.

Planners in England are required to mediate between pro- and anti-development positions. They are also continuing to negotiate the presumption in favour of economic growth framed within the NPPF's sustainable development rhetoric

(DCLG, 2012; Landscape Institute, 2012). Within this process there appears to be one main loser – the landscape. While it could be argued that the conservation of green spaces is a key component of supporting landscape sustainability, the preservation in stasis of green belts limits the ability of our urban-fringe landscapes to deliver greater socio-economic and ecological worth.

As discussed in Chapter 2, the NPPF maintains a clear focus for green belt policy in England. The rationale for this view includes safeguarding the countryside from encroachment by restricting sprawl and coalescence and preserving the special character of historic towns and countryside settings in England. Within this rationale, though, we can highlight problems with the use of emotive language and vague conceptualisations – for example, "special" – which, as noted previously, is highly subjective. The DCLG develops this view, stating that, once they have been designated, local authorities should aspire to improve the beneficial use of green belts by improving access and opportunities for both passive and active recreation. However, this must be balanced with a need to maintain biodiversity and to retain and/or enhance the landscape. It could therefore be argued that the UK government's proposals for green belts are simultaneously protectionist and expansionist in nature. Such a pluralistic assessment of green belts was proposed by Woods (2011), who suggested that, without a defined understanding of the complementary nature of town and country, green belts would remain an 'elusive' concept.

Chapter 3 discussed the impacts of green belts, including the tendency for development to "leapfrog" and continue beyond them. Peter Hall's work on the London green belt identified this issue, noting that, although the western extension of the protected area limited the expansion of London, it has led to the merging of towns in and around Reading at the edges of the green belt (Hall and Tewdwr-Jones, 2010). We might therefore argue that attempts to limit coalescence failed to appreciate the extent to which the evolving development context of places outside of the core cities would clash with protectionist views of the green belt.

We might also argue that the discussion of "special character" is, in reality, axiomatic, as it provides ample scope for people to misappropriate the word "special". While it would be counter-intuitive to state that some of the UK's towns and cities do not have unique characteristics, that particular phrasing provides opportunities for misuse or misinterpretation of the green belt. For example, the city of Ely lies outside the Cambridgeshire green belt but is framed by the cathedral and its views over the Fens. As a consequence, planning along specific sight lines in the green belt are restricted. However, developments in locations such as Histon which lie within the green belt are subject to additional scrutiny because they are deemed to support the rural character of the area. However, this small town, like many towns in the Cambridgeshire green belt, could be considered to be an "ordinary" town with no discernible specific features or characteristics. How such special features are discussed, whether positively or not, potentially places a large number of restrictions on the use of the landscape, many of which could actually be downplayed (Halfacree, 2006; Selman, 2009; Woods, 2011). The NPPF also states that local authorities should regard any application for the construction of new dwelling or buildings in the green belt as inappropriate. Maintaining the position introduced with Circular 42/55, there are certain categories where development in green belt is allowed – for example, if buildings are for agriculture or forestry (DCLG, 2012, pp. 20–21). In theory, affordable housing can be allowed in green belts, though previous research has found local authorities very reluctant to permit this (Best and Shucksmith, 2006). Managing green belt while also promoting affordability and meeting local needs can thus become compromised.

While ensuring that neither urban sprawl nor coalescence occurs, the protectionist rhetoric of green belt planning effectively isolates villages and communities from investment and the delivery of services such as high-speed broadband (Woods, 2011). It also highlights another issue, that of wealth and opportunity. In the nineteenth and twentieth centuries, rural communities were relatively secure in terms of employment and communal ties. However, as more young people moved to urban areas for employment or education, we started to see a counter-shift of people relocating to the countryside and green belt areas. A large percentage of these were affluent and, therefore, moved to areas within the green belt (a) because they were able to and (b) because they appreciated the aesthetic qualities of these areas (Hall andTewdwr-Jones, 2010). Furthermore, although the green belt offers the tranquillity of the countryside (which attracts people), there are rising conflicts over the need to develop infrastructure and service residential and tourist populations.

It might also be argued that migration to the countryside simply extends the notion of suburbia promoted in the early 1900s, which the green belt was aiming to control (Matless, 1998). The differences in the twenty-first century appears to be the form of transport used, the distance between the city and the countryside, and the expectations of new rural residents about what constitutes the countryside (Murdoch *et al.*, 2003). Discussions of peri-urban and rural development thus promote a view of a conceptual continuum of landscapes that melt into each other, in much the same way that Metroland and suburbia were discussed in the 1900s (Hunt, 2009). However, this places additional pressures on planners to accept that simple characterisation of green belts as sacrosanct to help shape their conservation or development objectives.

Such issues are now being re-explored to highlight the pluralism of how we should define green belts, questioning whether this is influencing a change in how people and policy interpret and interact with them (Cullingworth *et al.*, 2015). Resistance to change can be presented as illustrating a misplaced interpretation of the value of green belts based on elitist presumptions of landscape "value" developed in the eighteenth and nineteenth centuries (Hall, 2002). This implies that current thinking on the countryside is still being framed as a deliberate attempt to codify landscape values through a historical perspective, which translates into a nostalgic commodification of the green belts to maintain a specific conceptualisation of their composition (Jackson, 1980). We clearly see representations of this view within the NPPF and the campaigns of the CPRE. In reality, though, such a view is widely contested. For example, the long-term plans for the development of John Lennon International Airport in Liverpool include a freight/cargo terminal next to the River Mersey estuary, an area of the city's green belt. This land could be considered as urban-fringe scrubland, but, because of its classification, the airport authorities are subject to restrictions (John Lennon International Airport and Peel Holdings, 2006).

Understanding how the various assumptions relating to the green belt interact illustrates the difficulty in identifying who actually controls the development protection process; is it planners, developers, local communities or national politicians? The influences used to support decision-making in these areas also provide a useful insight into how debates are being framed locally as well as nationally.

In those local authorities which internalise development debates within the simplistic *yes–no* dichotomy, we are seeing, in one sense, an attempt to reinstate traditional or idealised views of the green belt in the UK. Often the process of decision-making is based on the utilisation of existing personal networks in council and community groups, which are used to garner support for the acceptance or refusal of development (Gallent, 2013). Where this occurs, we can see stand-offs between people who feel that green belts are sacrosanct and those who believe that landscapes need to provide more contemporary services and benefits for the local population. The use of personal influence, especially by campaign groups (including the CPRE), at both national and local level (Davies, 2014) could be regarded as limiting more legitimate attempts to improve the quality of life for all residents and workers in rural areas.

An ongoing issue in all discussions of development in the green belt is the growing prominence of campaigning organisations that are unwilling to engage in rational dialogue (Woods, 2011). Both the pro- and anti-development lobbies are guilty of this, while they simultaneously excel at engaging the media to promote the *cause célèbre* of green belt protection/development. The most successful of these organisations, the CPRE, has used a mixture of evocative imagery and emotive language to frame development as an unnecessary evil being forced upon rural communities (Sturzaker, 2010). One example is its 'Speak out for the countryside' campaign, which promotes three themes directly applicable to green belt debates:

(i) building on brownfield land first;
(ii) more housing – in the right places; and
(iii) a fair say for communities.

(CPRE, 2015b)

The campaign speaks of development in the 'right places', yet the campaign states that local communities have been 'left powerless' to shape local environments. In essence the CPRE is promoting a very specific form of localism that relies on a nostalgic interpretation of the countryside to protect green belts rather than addressing current issues – housing shortages, unemployment and underemployment, and a lack of infrastructure.

It could also be suggested that the countryside the CPRE claims to protect is, in reality, an imagined concept which fails to grasp the realities of a growing hybridity of economic, social and environmental factors in rural areas. In his research on landscape, memory and nostalgia, Lowenthal (1985) suggested that the past was different and perceived as idyllic. He goes further to debate how this view constrains more contemporary explorations of landscape development because it refuses to acknowledge that changes have occurred – changes which are now reshaping the utility of green belt functions. In response to such assertions, the CPRE has framed its discussions of the countryside as an "us vs. them" battle to *protect* important and valuable landscape resources (Sturzaker, 2010).

Conclusion

In contrast to the view of the CPRE, an alternative view can be put forward that moves away from the traditional protectionist positions to a hybridised development continuum (Cloke, 2006). If we take the view that the level of housing and infrastructure available in large swaths of peri-urban and rural Britain is incompatible with need, some commentators might suggest allowing further development in urban areas, particularly on brownfield sites, to direct these needs. This does not, however, address the underlying issue of service provision in green belt areas and the wider countryside. New homes and new or improved roads are needed, along with additional services (i.e., schools and doctors' surgeries), in both urban and rural locations. Therefore, a definitive "no" to development based on similar notions of Blake's "And did those feet" or the paintings of John Constable limits the ability of local authorities within the green belt to create functional communities for local people (Macnaghten and Urry, 1998; Schama, 1996).

It therefore seems illogical to continue maintaining views entrenched in nostalgic and potentially outdated ideologies of the green belt, given the evidence that their "value" may not be as sacrosanct as previously thought (Amati and Taylor, 2010). While we need to ensure that our socio-environmental relationships with the landscape are sustained, there may be less viability or relevance in revisiting the idealised views of our green belts in the twenty-first century. Promoting a more transparent process of development that offers clarity to both developers and the public of the form and location of contentious investments in the green belt would be one way of potentially addressing the entrenched positions of current debates. An increased level of engagement and discussion between stakeholders may be one way to reduce opposition to development in green belts, and there is some evidence that this works in rural areas more broadly (Sturzaker, 2011). Such a change, however, relies on stakeholders to hold somewhat flexible or malleable positions – a notion not readily seen in the majority of discussions about the green belt.

We might argue that there is a need to reframe development in the green belt as a test of relevance. For example, is a proposed development relevant to the long-term viability of a local community? Are opposition calls for protection based on contemporary or nostalgic assessments of the landscape, and are these relevant in the twenty-first century? Does the landscape character of the existing green belt place a legislative and social understanding on any proposed development? Furthermore, as we have suggested, perceptions of development are hybridised, reflecting both growth and conservation narratives. Therefore, if we acknowledge these shifts, we can potentially move the debate forward and away from existing simplistic positions (Cloke, 2006).

Within this chapter, Blake's "And did those feet" was quoted to frame the discussion of Britain's green belts. While some agencies, including the CPRE, argue that retaining this vision of a preserved landscape is essential, the current battleground of planning highlights an altogether different set of interpretations. This discussion, though, is not about choosing sides. Planning is about doing what is socially, ecologically and economically appropriate for the betterment of society within the legislative framework of policy and guidance. Planning professionals can therefore teach us that development in the green belt should be based on balanced negotiations and compromise, not necessarily on an entrenched view of a sacrosanct designation (Cullingworth *et al.*, 2015; Woods, 2011). It may, however, be possible to retain Blake's vision of England, but only if we can also plan for greater equality in access and rights to develop in the green belt. Although this may not be the most politically expedient stance, given the current house building targets coupled with the desire of people to move to the country, it seems logical that development must occur in *our Jerusalem*.

References

Abercrombie, P. (1945) *Greater London Plan* (London: HMSO).

Amati, M., and Taylor, L. (2010) From green belts to green infrastructure, *Planning Practice and Research*, 25(2): 143–55.

Amati, M., and Yokohari, M. (2006) Temporal changes and local variations in the functions of London's green belt, Landscape and Urban Planning, 75: 125–42.

Barlow, Sir M. (1940) *Report [of the] Royal Commission on the Distribution of the Industrial Population* (London: HMSO).

Best, R., and Shucksmith, M. (2006) *Homes for Rural Communities: Report of the Joseph Rowntree Foundation Rural Housing Policy Forum* (York: Joseph Rowntree Foundation).

Brace, C. (1999) Finding England everywhere: regional identity and the construction of national identity, 1890–1940, *Cultural Geographies*, 6(1): 90–109.

Cheshire, P. (2013) Greenbelt myth is the driving force behind housing crisis, *The Conversation*, 9 September, http://theconversation.com/greenbelt-myth-is-the-driving-force-behind-housing-crisis-17802.

Cheshire, P. (2014) Turning houses into gold: the failure of British planning, *CentrePiece: the Magazine of the Centre for Economic Performance*, 19(1): 14–18.

Clark, R. (2015) Fringe benefits, *Sunday Times*, 4 October, pp. 8–9.

Cloke, P. (2006) Conceptualizing rurality, in P. Cloke, T. Marsden and P. Mooney (eds), *Handbook of Rural Studies* (London: Sage), pp. 18–27.

Cloke, P., and Little, J. (1997) *Contested Countryside Cultures: Rurality and Socio-Cultural Marginalisation* (Abingdon: Routledge).

Cosgrove, D. (1985) Prospect, perspective and the evolution of the landscape idea, *Transactions of the Institute of British Geographers*, 10(1): 45–62.

Countryside Agency and Groundwork (2005) *The Countryside in and around Towns: A Vision for Connecting Town and County in the Pursuit of Sustainable Development* (Wetherby: Countryside Agency & Groundwork).

Countryside Commission (1994) *Countryside Character Programme Information (CCP 472)* (Cheltenham: Countryside Commission).

CPRE (2015a) 60th anniversary poll shows clear support for green belt, www.cpre.org. uk/media-centre/latest-news-releases/item/4033-60th-anniversary-poll-shows-clear-support-for-green-belt.

CPRE (2015b) What we've achieved with our charter to save the countryside, Available www.cpre.org.uk/what-we-do/housing-and-planning/planning/update/item/3989-what-we-have-achieved-with-our-charter-to-save-the-countryside.

CPRE (2016) *Green Belt under Siege: 2016*, www.cpre.org.uk/resources/housing-and-planning/green-belts/item/4276-green-belt-under-siege-2016.

CPRE and Natural England (2010) *Green Belts: A Greener Future* (London: CPRE & Natural England).

Cullingworth, B., Nadin, V., Hart, T., Davoudi, S., Pendlebury, J., Vigar, G., Webb, D., and Townshend, T. (2015) *Town and Country Planning in the UK* (15th ed., London: Routledge).

Davies, C., Macfarlane, R., McGloin, C., and Roe, M. (2006) *Green Infrastructure Planning Guide* (Annfield Plain: North East Community Forest).

Davies, H. (2014) Whiston campaigners fighting to save town's green belt, *Liverpool Echo*, 23 October, www.liverpoolecho.co.uk/news/liverpool-news/whiston-campaigners-fighting-save-towns-7984702.

DCLG (2012) *National Planning Policy Framework* (London: Department for Communities and Local Government).

DCLG (2015) *Local Planning Authority Green Belt: England 2014/15* (London: Department for Communities and Local Government).

England's Community Forests (2004) *Quality of Place, Quality of Life* (Newcastle: England's Community Forests).

Evans, C., and Freestone, R. (2010) From green belt to green web: regional open space planning in Sydney, 1948–1963, *Planning Practice and Research*, 25(2): 223–40.

Gallent, N. (2013) Re-connecting 'people and planning': parish plans and the English localism agenda, *Town Planning Review*, 84(3): 371–96.

Gallent, N., Andersson, J., and Bianconi, M. (2006) *Planning on the Edge: The Context for Planning at the Rural–Urban Fringe* (London: Routledge).

Gallent, N., Hamiduddin, I., Juntii, M., Kidd, S., and Shaw, D. (2015) *Introducing Rural Planning: Economies, Communities and Landscapes* (2nd ed., Abingdon: Routledge).

Gallent, N., Shoard, M., Andersson, J., Oades, R., and Tudor, C. (2004) Inspiring England's urban fringes: multi-functionality and planning, *Local Environment*, 9(3): 217–33.

Gilg, A. (1996) *Countryside Planning: The First Half Century* (2nd ed., London: Routledge).

Halfacree, K. (2006) Rural space: constructing a three-fold architecture, in P. Cloke, T. Marsden and P. Mooney (eds), *Handbook of Rural Studies* (London: Sage), pp. 44–62.

Hall, P. (2002) *Cities of Tomorrow: An Intellectual History of Urban Planning and Design in the Twentieth Century* (Oxford: Blackwell).

Hall, P., and Tewdwr-Jones, M. (2010) *Urban and Regional Planning* (Abingdon: Routledge).

Healey, P., and Shaw, T. (1994) Changing meanings of 'environment' in the British planning system, *Transactions of the Institute of British Geographers*, 19(4): 425–38.

Herrington, S. (2009) *On Landscapes* (New York: Routledge).

House of Commons South East Regional Committee (2010) *Housing in the South East – First Report of Session 2009–10: Report, together with Formal Minutes, Oral and Written Evidence* (London: The Stationery Office).

Howard, E. (1898) *To-Morrow: A Peaceful Path to Real Reform* (London: Swann Sonnenschein).

Hunt, T. (2009) The suburbs are derided by snobs, yet they offer hope for our future, *The Guardian*, 19 July, www.theguardian.com/commentisfree/2009/jul/19/suburbs-snobbery.

Jackson, J. B. (1980) *The Necessity of Ruins and Other Topics* (Amherst: University of Massachusetts Press).

John Lennon International Airport and Peel Holdings (2006) *John Lennon International Airport: Airport Master Plan to 2030* (Liverpool: John Lennon International Airport & Peel Holdings).

Jongman, R., and Pungetti, G. (2004) *Ecological Networks and Greenways: Concept, Design and Implementation* (Cambridge: Cambridge University Press).

Kaplan, R., and Kaplan, S. (1989) *The Experience of Nature: A Psychological Perspective* (New York: Cambridge University Press).

Landscape Institute (2012) National Planning Policy Framework – has the government listened?, Press release, 29 March, www.landscapeinstitute.org/PDF/Contribute/LINPPFresponse.pdf.

Little, C. (1990) *Greenways for America* (Baltimore: John Hopkins University Press).

Lowenthal, D. (1985) *The Past is a Foreign Country* (Cambridge: Cambridge University Press).

Macnaghten, P., and Urry, J. (1998) *Contested Natures* (London: Sage).

Matless, D. (1998) *Landscape and Englishness* (London: Reaktion Books).

Mell, I. C. (2010) Green infrastructure: concepts, perceptions and its use in spatial planning, Unpublished PhD thesis, Newcastle University.

Mell, I. C. (2016) *Global Green Infrastructure: Lessons for Successful Policy-Making, Investment and Management* (Abingdon: Routledge).

Mell, I. C., Henneberry, J., Hehl-Lange, S., and Keskin, B. (2013) Promoting urban greening: valuing the development of green infrastructure investments in the urban core of Manchester, UK, *Urban Forestry & Urban Greening*, 12(3): 296–306.

Mormont, M. (1990) Who is rural? Or, how to be rural: towards a sociology of the rural, in T. Marsden, P. Lowe and S. Whatmore (eds), *Rural Restructuring: Global Processes and their Local Response* (London: David Fulton), pp. 21–44.

Morrison, N. (2010) A green belt under pressure: the case of Cambridge, England, *Planning Practice and Research*, 25(2): 157–81.

Murdoch, J., Lowe, P., Ward, N., and Marsden, T. (2003) *The Differentiated Countryside* (New York: Routledge).

Nassauer, J. (1995) Culture and changing landscape structure, *Landscape Ecology*, 10(4): 229–37.

National Trust (2015) *AONBs and Development* (Rotherham: National Trust).

Natural England (2010) *England's Statutory Landscape Designations: A Practical Guide to your Duty of Regard* (Peterborough: Natural England).

Papworth, T. (2015) *The Green Noose: An Analysis of Green Belts and Proposals for Reform* (London: ASI (Research) Ltd).

Pauleit, S., Ennos, R., and Golding, Y. (2005) Modeling the environmental impacts of urban land use and land cover change – a study in Merseyside, UK, *Landscape and Urban Planning*, 71(2–4): 295–310.

Pepper, D. (1996) *Modern Environmentalism: An Introduction* (Abingdon: Routledge).

Quilty-Harper, C., Palmer, D., and Oliver, M. (2012) Interactive map: England's green belt, *The Telegraph*, 28 November, www.telegraph.co.uk/news/earth/greenpolitics/planning/9708387/Interactive-map-Englands-green-belt.html.

Richards, F., and Satsangi, M. (2004) Importing a problem? Affordable housing in Britain's national parks, *Planning, Practice and Research*, 19(3): 251–66.

Roe, M., and Mell, I. C. (2013) Negotiating value and priorities: evaluating the demands of green infrastructure development, *Journal of Environmental Planning and Management*, 56(5): 37–41.

Roe, M. H. and Taylor, K. (eds) (2014) *New Cultural Landscapes* (London: Routledge).

Roe, M. H., Selman, P. H., Mell, I. C., Jones, C., and Swanwick, C. (2009) *Establishment of a Baseline for, and Monitoring of the Impact of, the European Landscape Convention in the UK*, Defra Contract No. CR0401 (Comp. Code: WC0802), Bristol.

Satsangi, M., and Dunmore, K. (2003) The planning system and the provision of affordable housing in rural Britain: a comparison of the Scottish and English experience, *Housing Studies*, 18(2): 201–17.

Schama, S. (1996) *Landscape & Memory* (London: Fontana).

Schively, C. (2007) Understanding the NIMBY and LULU phenomena: reassessing our knowledge base and informing future research, *Journal of Planning Literature*, 21(3): 255–66.

Scottish Government (2014) *Scottish Planning Policy* (Edinburgh: Scottish Government).

Scottish Office (1996) *Natural Heritage Designations Review: Discussion Paper* (Edinburgh: Scottish Office).

Selman, P. (2006) *Planning at the Landscape Scale* (Abingdon: Routledge).

Selman, P. (2009) Planning for landscape multifunctionality, *Sustainability: Science, Practice and Policy*, 5(2): 45–52.

Smith, L. (2015) *Planning for Housing*, House of Commons Library Briefing Paper no. 03741 (London: House of Commons).

South Yorkshire Forest Partnership and Sheffield City Council (2012) *The VALUE Project: The Final Report* (Sheffield: South Yorkshire Forest Partnership & Sheffield City Council).

Spanò, M., DeBellis, Y., Sanesi, G., and Lafortezza, R. (2015) *Green Surge: Milan, Italy – Case Study City Portrait: Part of a GREEN SURGE Study on Urban Green Infrastructure Planning and Governance in 20 European Cities* (Copenhagen: University of Copenhagen).

Spirn, A. W. (1998) *The Language of Landscape* (New Haven, CT: Yale University Press).

Sturzaker, J. (2010) The exercise of power to limit the development of new housing in the English countryside, *Environment and Planning A*, 42(4): 1001–16.

Sturzaker, J. (2011) Can community empowerment reduce opposition to housing? Evidence from rural England, *Planning Practice and Research*, 26(5): 555–70.

Thomas, K., and Littlewood, S. (2010) From green belts to green infrastructure? The evolution of a new concept in the emerging soft governance of spatial strategies, *Planning Practice and Research*, 25(2): 203–22.

Tuan, Y. (1990) *Topophilia: A Study of Environmental Perceptions, Attitudes, and Values* (New York: Columbia University Press).

Tudor, C. (2014) *An Approach to Landscape Character Assessment* (Peterborough: Natural England).

Tyrväinen, L., and Väänänen, H. (1998) The economic value of urban forest amenities: an application of the contingent valuation method, *Landscape and Urban Planning*, 43(1–3): 105–18.

UNEP–WCMC (2011) *UK National Ecosystem Assessment: Understanding Nature's Value to Society. Synthesis of Key Findings* (Cambridge: United Nations Environment Programme World Conservation Monitoring Centre).

Wolsink, M. (2006) Invalid theory impedes our understanding: a critique on the persistence of the language of NIMBY, *Transactions of the Institute of British Geographers*, 31(1): 85–91.

Woods, M. (2011) *Rural* (London: Routledge).

5　Green belts

International case studies

Introduction

Since their introduction in the UK in the 1930s, we have seen a growing use of "green belts", or similar policies under different names, around the world. As with any other aspect of planning, the application of green belt principles differs depending on the political and socio-economic context of a given location. While many international examples of green belts retain a comparable spatial form, the manner in which they are planned for and managed is more variable, as are their purposes. This chapter explores some different examples of green belts and examines how the development arenas of different countries influence their use, management and valuation. Each of the case studies discusses the nuances and reflects on both historical and current debates of the utility of green belts as a form of landscape and urban planning. The selection of examples does not profess to be exhaustive; rather, it is a snapshot of those which have been discussed in the academic and practitioner literature. Finally, the main successes, themes and barriers to implementation are brought together to summarise key overarching points.

Asia

Seoul (and South Korea)

The Seoul green belt has been extensively studied, the other green belts in South Korea less so, though most were introduced at the same time. Inspired by the Greater London Plan, the Korean Planners Association proposed a green belt around Seoul in 1964 (Bae, 1998). Between 1950 and 1975, Seoul was the most rapidly developing city in the world, with an average annual growth rate of 7.6 per cent (Bengston and Youn, 2006) expanding its population from 1 to 6.8 million people. Seoul is near the demilitarised zone between North and South Korea, and the close proximity of so many people to the border was a concern to the South Korean government. Bengston and Youn estimate that, in 1970, 40 per cent of the South Korean population was within reach of an artillery attack from the North. In order to limit the growth of Seoul – and thirteen other cities – green belts, or restricted development zones, were introduced between 1971 and 1973.

Seoul's green belt is approximately 10 kilometres wide, starting 15 kilometres from Seoul's central business district. In 1976, shortly after its introduction, it had an area of 1,500 square kilometres, approximately 13 per cent of the Seoul Metropolitan Area, and contained only 1.7 per cent of the population (Bengston and Youn, 2006). In total, the fourteen green belts in South Korea covered 5.4 per cent of the country's land area in 1998; 80 per cent of this was privately owned, 60 per cent consisted of mountains and forests, 27 per cent was agricultural land, and 3.6 per cent was residential (Bae, 1998).

Beyond the need to limit development too close to the demilitarised zone, Bae (1998) identified six further objectives for Seoul's green belt:

1   to 'eliminate illegal suburban shantytowns' (ibid., p. 483);
2   to discourage urban sprawl;
3   to harmonise 'metropolitan growth by controlling growth within the Greenbelt and promoting suburban and exurban growth beyond the Greenbelt' (ibid., p. 484);
4   to control land speculation;
5   to protect agricultural land;
6   to protect environmental and natural resources.

As in the UK, as the Korean economy has grown and environmental awareness has become more prominent, the environmental rationale for the Seoul green belt has become more important. Similarly, the density of the population in the capital city has meant more and more people need some relief from urban life; the 60 per cent of the Seoul green belt comprising mountains and forests is heavily used for recreation (Bengston and Youn, 2006).

Some have argued that the hasty implementation of restricted development zones and the lack of public input has meant they were flawed from the outset (Bengston and Youn, 2006; Kim and Kim, 2008), and there have been doubts as to whether they were required in some of the smaller cities beyond Seoul (Choe, 2004). However, as at that time South Korea was ruled by a military regime, public opposition to the green belt was prohibited (Lee and Linneman, 1998; Bengston and Youn, 2006). This changed in 1988 with the transition to democracy (Bengston and Youn, 2006), and landowners lobbied against the green belt reducing the value of their assets.

In 1997 a new president (Kim Dae-jung) was elected who had campaigned, among other things, on reviewing and reforming the green belt. The following year a reform committee was set up, which recommended the removal of green belt around smaller cities that were under less pressure for development, the revision of green belt boundaries, and the introduction of compensation for landowners. A lobby group was set up to oppose these recommendations. The TCPA was asked to comment on the committee's report, resulting in 'divergent views about whether it supported the reforms' (Bengston and Youn, 2006, p. 9). In 1999 the government followed many of the recommendations and unilaterally announced a new restricted development zone policy. Green belts around seven cities were removed and land was redesignated as 'conservation-green' or 'natural-green' areas. In the other seven cities, including Seoul, green belts have been redelineated through environmental assessment and graded 1 to 5, with grades 4 and 5 subject to further review to consider the varying of restrictions (Choe, 2004). Following these changes, some land around Seoul was released through the fifteen-year metropolitan plan, though opposition continued in the media.

A number of authors (cf. Amati, 2008) have undertaken studies of green belts in South Korea and found that some of their impacts are similar to those of the UK green belts, perhaps because they have been used for a similar purpose – *controlling* urban growth, not, as in cases we will discuss below, *managing* that growth. It appears that green belts have limited urban sprawl in South Korea (Bengston and Youn, 2006), but with a number of costs. Firstly, urban density within the green belts has increased substantially (Bae, 1998; Lee, 1999; Choe, 2004). Secondly, leapfrogging has occurred (Bae, 1998; Bengston and Youn, 2006; Kim and Kim, 2008). Thirdly, there has been a housing shortage, resulting in higher prices in locations outside of the green belt (Kim and Kim, 2008).

However, the green belt 'remains entrenched as the key element in Korea's land use controls' (Bae, 1998, p. 479), despite being considered by some as 'inefficient and inequitable' (Kim and Kim, 2000, p. 1158). The reasons for the public's continued support again appear similar to those in the UK: strong media backing; differences between public perception of the quality of green belts and the reality, with the green belt often used 'as a dumping ground for locally undesirable land uses' (Bae, 1998, p. 485); and a failure to understand the costs of green belts (Kim and Kim, 2000).

China – urban sprawl control vs. land release

China's population has urbanised extremely rapidly: whereas in 1950 only 12 per cent of the population lived in urban areas, now it is 50 per cent (United Nations Department of Economic and Social Affairs, 2014). This massive change has inevitably led to pressure on those urban areas, and sprawl has been a major issue. Unlike the sprawl found in the UK or other Western countries, it is not typically low-density suburbia; rather, both migration and economic growth mean it can be of high density (Yang and Jinxing, 2007). The Chinese government has taken steps to control it, primarily perhaps to protect agricultural land in order to maintain food supplies (Zhang, 2000; Zhao *et al.*, 2010), by such means as limiting the size of city populations, restricting the conversion of agricultural land to other uses, imposing urban density targets, and introducing green belts (Zhang, 2000; Yang and Jinxing, 2007; Zhao *et al.*, 2010; Zhao, 2011). Green belts have been introduced in various large cities, including Beijing and Shanghai, as we will discuss below.

Before we do so, it is necessary briefly to explain the impact that the market reforms of the last twenty-five years have had on land supply. In the post-reform system, the ownership of land-use rights can be sold, but land itself cannot. In rural areas, land is collectively owned, but in urban areas it is owned by the state.

The state does not, though, own the buildings or other structures on that land (Tian and Ma, 2009). Power and financial responsibility has been decentralised to the municipalities, and beyond, to district and county governments within municipalities (Zhao, 2011). Local governments can make their own decisions on economic development, which they are expected to encourage. This means they are inclined to attempt to expand their urban areas to facilitate more development, which of course can run counter to attempts to limit urban sprawl (Zhang, 2000; Zhao *et al.*, 2010). There is also a model of what has been called 'local state corporatism' (Zhao, 2011, p. 97), wherein local government officials sit on boards of directors of local enterprises, which can lead to conflicts of interest.

So how do these issues affect urban growth in China?

Beijing

A green belt around Beijing was first proposed many years before the market reforms, in 1958 (Yang and Jinxing, 2007). It was part of the general plan for the city of that year, which proposed that the city be organised on the basis of functional zones, separated with green spaces for use as agriculture and horticulture. Those green spaces, 314 square kilometres in area, would also have been used to 'help control the growth of the planned city area and separate the city area from the rural area' (ibid., p. 289). The second general plan of 1983 reaffirmed the principles of green belts laid down in the 1958 plan but was adapted to reflect the unregulated development that had taken place, which had resulted in an expansion of the built-up area from 84 to 371 square kilometres. The green belt was thus reduced in size to 260 square kilometres. It was proposed it should consist of five forest parks and nine 'restricted areas', to be used to separate the urban core from outer suburbs, which were to be planned on the basis of 'decentralised concentration' (Zhao *et al.*, 2010, p. 148).

Success was mixed. By 1993, the built-up area had further increased to 397 square kilometres, so a third general plan was prepared, proposing a second green belt (of 1556 square kilometres) and reducing the size of the first again to 240 square kilometres. Finally, the fourth general plan, produced in 2004, maintained the size of the first green belt and increased the size of the second to 1620 square kilometres. The first green belt was also redesignated as having functions beyond preventing coalescence, including 'recreational, visual and ecological benefits' (Yang and Jinxing, 2007, p. 290).

The second green belt was expected to be 10 per cent 'greenspace' by 2008, 50 per cent by 2020. It is interesting to note that 'greenspace' in this context does not include agricultural land. The Beijing municipal government focused on converting agricultural land into greenspace by turning it into parks and planting trees; quotas were established for tree planting which were passed down through the chain of local administration, to district, town, ward and village governments.

Despite the green belts and the attempts noted above by the central state to limit urban sprawl, the built-up area of Beijing increased sixfold between 1979 to 2004, from 180 to 1086 square kilometres (Zhao *et al.*, 2010). So have attempts to control urban growth had any effect? The evidence suggests that these have managed to increase the compactness of the city and focus the largest amount of growth in the peripheral "constellations". However, development has taken place everywhere across the city, including in the urban fringe, often as a result of 'illegal' employment land designations by district, county or town governments (ibid.; Zhao, 2011). Beyond this, Yang and Jinxing (2007, p. 293) argue that, for three reasons, the first green belt failed to separate the urban core from peripheral developments: firstly, it failed to accurately predict the growth in the city's population; secondly, the plan was top-down and did not involve 'key stakeholders'; and, thirdly, the plan overemphasized the function of separation and neglected other functions. The similarity to the UK experience here is striking. Yang and Jinxing (ibid.) conclude that the Chinese government had a 'false belief' that it could control urban growth, as it had reformed its economy but maintained its old central planning system. As in the UK in the post-war years, the market dictated where much development took place, and the planning system was ill-suited to dealing with this.

Shanghai

Shanghai is China's largest city and economic hub, so the development pressures placed upon its landscape are substantial. It is home to over 23 million people who reside within a urban footprint of approximately 6314 square kilometres (Wang *et al.*, 2014). Only since 1994, and specifically from 1996 onwards, has green space increased significantly in the city and its surrounding areas (Ren *et al.*, 2003). Moreover, before 1994 Shanghai had approximately 1.42 square kilometres of green belt land (by 1996 it had risen to 7.85 square kilometres). However, after 1996 the city authorities proposed to invest in a further 580,000 hectares of new or redeveloped green spaces and set themselves a deadline of 2030 to achieve this goal (Yang and Jinxing, 2007). The expansion of the green belt has been part of a wider programme of urban greening undertaken by the city's government that has included investment in linear urban features, gardens and parks, some of which are also called *green belt parks* (Wang *et al.*, 2014).

The increased focus placed on the environment by Shanghai's municipal government is linked directly to the need to contain urban sprawl and protect the limited green space resource base. This reflects what Wang *et al.* (2014) referred to as addressing the mixture of the rural and weak urban landscape. To increase the proportion of green space, Shanghai's government proposed the creation of a forested green belt 97 kilometres long and 0.5 kilometres wide around the city to limit ongoing expansion and coalescence with other urban areas (Ren *et al.*, 2003; Wang *et al.*, 2014). This circular corridor would form a 'long vine with melons', where the melons would be new parks, similar to the links, hubs and nodes principles of landscape ecology (Jongman and Pungetti, 2004). By 2005, 50 per cent of the green belt had been completed, which by 2009 had grown to 42 square kilometres. The cost of the investment was estimated to be approximately 13.9 billion RMB, roughly US$2.1 billion or £1.45 billion (Wang *et al.*, 2014).

One further use of green belts in Shanghai is interesting. The term "green belt" has been used in urban Shanghai, and specifically in the financial area of Pudong, to include urban parks (as with Beijing). Although the spatial footprint, location and linear/circular nature of these sites, such as Luijazui Central Green Belt park, differ substantially from traditional interpretations of green belts, planners and landscape architects are using the terminology to promote the visibility of these new resources (ibid.).

India

The development and management of green belt areas in India varies depending on local ecological conditions, urban growth, and the real, and perceived, authority of government at various levels to manage the environment. Nationally, the Ministry of Environment and Forests (MoEF) has produced advisory policy which requires state and local government to conserve landscape resources, referring specifically of large areas of green and open space and/or green belts. The guidelines developed from the MoEF (Government of India, 1980) include:

1 no forest areas within a green belt shall be converted to industry;
2 no agricultural areas within a green belt shall be converted to industry;
3 any industrial development shall be concealed from sight by green belt resources (normally trees);
4 development will include waste water treatment which will be used to improve the green belt;
5 a 1 km green belt buffer will be developed between industrial sites with at least 0.5km between industry that releases odour and other developments;
6 mandatory tree planting on roads to form green belts.

However, despite the advocacy of both the MoEF and the National Green Tribunal of green belt protection, there is no adopted national-level management policy (Jain, 2014; *Times of India*, 2015). In 1994 the MoEF extended its policies in an attempt to integrate a more structured process of monitoring and evaluation into green belt protection. This led to the development of Environmental Impact Assessment strategies being included in guidance (revised in 2006), which outlined the ecological and socio-economic value of green belt designations. The MoEF also proposed mechanisms to protect green belt from development in the long term. This, however, led to a diverse range of state- and city-government-led policies which identify green belts as locations for protection in some places and for development in others. Each city in India can, therefore, take a nuanced view of how they manage growth. Although the National Green Tribunal and the MoEF have developed guidelines for what housing, commercial and infrastructure development objectives should be adhered to, these are often downplayed or not enforced.

Furthermore, India has at least two interpretations of *what* a green belt *is*. The first is similar to that in the UK, where tracts of land in the urban–rural interface are designated to ensure that coalescence or overexploitation doesn't occur.

The second promotes investment in linear features associated with industrial and infrastructure development to form buffers between these sites and other urban areas. The latter type are smaller and used more frequently in urban locations, involving tree planting as the main approach to green belt development (Press Trust of India, 2015). This has been scaled up from local street-level investments to the national level, as 1.1 million kilometres of road building has taken place with 2.1 billion trees having been planted. Green belts are also considered to have a direct influence on climate change management, focusing on the alleviation of air pollution, temperature and rainfall infiltration. Moderation of urban environmental stresses is therefore of critical importance both to the promotion of a liveable and prosperous environment and to meet local/national pollution targets (*India Today*, 2010).

New Delhi and the National Capital Region (NCR)

Following India's independence from Britain, the political structure changed to reflect the regional governance of different states. New Delhi is one such area where the city has to interact with the National Capital Region (NCR), the metropolitan region within which the city sits, as the NCR supplies it with social, economic and ecological resources. However, development in the neighbouring states of Haryana and Uttar Pradesh, as well as in New Delhi itself, has had a significant impact over time on the nature of the area's resource base. For example, the forest cover of the NCR is approximately 6.2 per cent, compared with 21.05 per cent nationally (Dash, 2014; Jayaseelan, 2015). Furthermore, there has been extensive tree felling to allow for the development of highways and housing, and the NCR Planning Board estimates that 32,769 hectares of green/green belt land and 1,464 hectares of water resources were lost to development between 1999 and 2012. One example of an area attempting to address this issue is the affluent district of Saket in southern Delhi, which is engaged in developing a green belt consisting of over 1,500 trees along the Mehrauli Badarpur Highway; this will form a buffer zone some 13 metres wide and 13 kilometres long (Kumar, 2015).

Historically, Delhi has a good record of managing its green infrastructure. The Delhi Ridge, a hilly area to the east of the urban core, has been managed through long-term afforestation, and, with the Lutyens development of New Delhi under the colonial administration after 1912, there was a concerted effort to create a "preserved forest". This helps both to retain the existing green spaces in and around the Ridge area but also to extend its boundaries to the north, west and south of the city. The Ridge is considered to be the 'lungs of the city' and has been identified within the Delhi Development Plan (Delhi Development Authority, 2007) as an area that should be protected. This has not, however, deterred developers from undertaking illegal building, and the physical scale (in square hectares) of the Ridge area is widely contested: opposing views state it is growing or shrinking as a result of protection and/or development (Chaudhry *et al.*, 2011; Mell, 2016).

In the wider NCR there are ongoing campaigns in the municipalities of Gurgaon, Dwarka and Ghaziabad to halt the illegal concretisation of green spaces and green belt lands (Chatterji, 2013; Narain, 2009). For instance, the Ghaziabad Development Authority Master Plan–2021 called for 15.97 per cent of land to be green belt or green space development, but ongoing growth continues to decrease this figure (Ghaziabad Development Authority, 2005). Moreover, in areas with designated green belts, such as the area known as Sector 8 (Raj Nagar) on the eastern fringe of the NCR, protected land is being exploited or developed with little or no concern for the loss of landscape functionality (*Times of India*, 2015). As a consequence of the perceived lack of green belt protection, there have been calls within the NCR from government and environmental advocates to increase the proportion of green space in Haryana by 400 per cent and in Uttar Pradesh by 500 per cent to respond to the loss of land to development. But it has proved difficult to ensure that the state governments of Haryana and Uttar Pradesh meet these targets, especially when New Delhi itself is not considered to be engaged in successful management of the city's green belt areas (Dash, 2014).

Bengaluru (Bangalore)

As with green belt protection in New Delhi, the management of the landscape, and in particular green spaces, in Bengaluru has been heavily influenced by urban development. The population of Bengaluru expanded from approximately 400,000 in 1941 to more than 8.5 million in 2012, which has led to substantial changes in the spatial footprint of the city and the subsequent reactions of local government to meeting the needs of its population. To provide housing, transport infrastructure and employment opportunities, Bengaluru expanded rapidly, most noticeably with the development in 2008 of a new airport 40 kilometres from the city centre, as the previous one had been surrounded by built infrastructure (Sudha and Ravindranath, 2000). One of the consequences of this has been the decreased spatial scale and functionality of the green belt. Such significant change also led to the Bengaluru's "garden city" branding becoming contested (Nagendra *et al.*, 2012).

Now, though, there is a concerted effort, coordinated through the Bengaluru Metropolitan Region Development Authority (BMRDA), to re-establish the garden-city mindset through investment in green belts and city-scale green spaces. In its latest policy announcements, the BMRDA has stated that it aims to maintain the city's green belt (approximately 742 square kilometres in size) in the long term (Bangalore Development Authority, 2005). This is being implemented partly through a city-scale tree planting project, which will see 400 acres (160 hectares) of urban forest built by the Army Supply Corps using 180,000 saplings. The BMRDA is also gathering further evidence of the climatic benefits of green belts in order to prevent the conversion of land from green belt to urban infrastructure (Nagendra and Gopal, 2010).

Ahmedabad

Ahmedabad, as the economic hub of the state of Gujarat, is a further example of a city whose green belt is subject to growth pressures. As the city expanded southwards and westwards across the Sabarmati River, the Ahmedabad Urban

Development Authority and the Ahmedabad Metropolitan Council, both local government planning agencies, developed a green belt designation within the first intention of the Ahmedabad Development Plan (Ahmedabad Urban Development Authority, 2013). The green belt was proposed as a buffer zone to limit the expansion of the urban core into agricultural land and was a key policy in the city's development plans throughout the 1970s (Mell, 2016). However, as the city's economic position increased both in the region and nationally, there have been calls to facilitate development opportunities for housing, transport and commercial infrastructure. As a consequence of these ongoing pressures (and the associated political influence), both planning agencies proposed to rescind the green belt designation in the latest development plan (Adhvaryu, 2011; Mathur, 2012). This was viewed as a retrograde step by many local environmental and academic stakeholders, who argue that such changes would limit the protection afforded to environmental resources in the city, as developers would no longer be constrained by the physical (or hypothetical) green belt boundary.

Australia

Unlike the green belt around Sydney, which has been undermined by ongoing urban encroachment, the designations in Adelaide and Melbourne remain functional and sizable resources. Although both these have been modified in terms of their location, size and accessibility since they were legislated for, there is an ongoing and positive commentary within the evaluations of each (Buxton and Goodman, 2008). As a result of strong legal and city government steering, both cities have retained their green belts despite the pressures being placed on the spatial footprint of their built environments, and both could be identified as examples of good practice for investment in and management of the landscape. Here we focus on Melbourne.

Melbourne

The development of the green network for Melbourne was conceived in four stages starting in 1929. From 1929 to 1954, a process of strategic planning was established that would have included the provision for a city-scale green belt. However, only from 1954 onwards, and specifically from 1968, did the current spatial form of Melbourne's green belt take shape (Buxton and Goodman, 2008). As the city's urban form was modified, in part through population change and commercial needs, Rushman (1969) highlighted that the green wedges identified as Melbourne's "green belt" were the lands remaining following major urban extension/expansion in the 1950s. The promotion of "wedges", however, was a forward-thinking policy to protect the environment using a city-wide green network. These green wedges cover approximately 6478 square kilometres, three times more land than within the urban growth boundary (UGB), and are distributed in a series of distinctive areas offering a mixture of multifunctional, agricultural and non-urban activities to be undertaken in urban-fringe areas.

Melbourne's green belt could be considered to take inspiration from the historical applications in both London and Copenhagen (Buxton and Goodman, 2008).

Green wedges were first conceived in 1966–8, legislated for in 1971, and protected under the State of Victoria Planning & Environment Act (1987, amendments 2003) (State of Victoria, 2002). According to Mees (2003), legislation for the city's green wedges formed part of the revision to the local growth plan. Planners used the linear form of the city (and its transport corridors) to link these developments with the green wedges, making use of radials and corridors to create a connected network.

Buxton (2014) helped define the direction of planning in Melbourne over the last forty years and the location of green belts within this narrative. Despite initial objections to seemingly restrictive zones, these have been set aside over time, as the green wedges have become part of the long-term planning of the city's urban framework (similar in that sense to the UK) (Buxton and Goodman, 2008). Furthermore, as the scale of development increases, the ecological and socio-economic value of retention likewise goes up, because the wedges have provided a positive development landscape for the city. Unfortunately, successive rezoning has led to changes in their extent, with areas being converted from green space to housing and/or infrastructure. While more recent policy, *Melbourne 2030* (State of Victoria, 2002), extended the protection given to the wedges, proposals for new housing and/or transport are being discussed as possible developments which could lead to the release/rezoning of some of them.

The *Melbourne 2030* strategy sets out how to plan to contain growth within the UGB in order to protect the green wedges. This, along with state legislation, offers a strong legal backing for their management, although some commentators have claimed that the forecasts of growth will require the boundary to be re-evaluated – i.e., within the *Melbourne @ 5 Million* document, which forecasts much higher housing requirement (284,000 rather than 134,000 homes) (State of Victoria, 2008). In an attempt to manage the competing development objectives currently being witnessed in Melbourne, a metropolitan structure of governance has been created. The green belt is managed through a metropolitan fringe policy that brings together eight municipal authorities to protect green wedges, all of which sets the legal framework for their protection within the Strategic Growth Plan (Buxton and Goodman, 2008; State of Victoria, 2002). The use of such a collective and engaged process allows discussions to occur within a more integrated setting, limiting the issues of 'institutional fragmentation [that] is fatal to Green Belts' (Buxton and Goodman, 2008, p.76).

The Melbourne green belt could thus be viewed as successful, as, despite objections from landowners and overtures from developers, it retains the majority of the ecological resource base. Its spatial form is also beneficial as it provides for a range of uses both in the urban core and in more rural areas (Buxton and Goodman, 2008).

Europe

There are green belts, or approaches to managing land similar to green belts, in various cities across Europe, including Barcelona, Budapest, Vienna and Berlin

(Kühn, 2003). In this section we look in more detail at four of them, the first of which is not a green belt at all – it in fact inverts the concept, putting the green at the *centre* of the urban area.

The Netherlands

The Green Heart in the Netherlands is part of the *Randstad*, or "rim city". The latter has existed as a term since the 1930s and describes the urban areas in the west of the Netherlands which take in Rotterdam, Amsterdam, The Hague/ Delft, Dordrecht and Utrecht. These urban areas, although not contiguous, form more or less a horseshoe shape around a large relatively undeveloped space – the Green Heart. In the 1950s this space was considered worthy of being maintained in this undeveloped way 'because the proposed urban development in the Randstad required open space on a concomitant scale' (van der Valk and Faludi, 1997, p. 61). Van der Valk and Faludi note that planners used the example of the green belt around London to support the need for this kind of protected open space.

The 1970s saw the strengthening of protection for the Green Heart, with national and provincial plans limiting development therein. In 1990 a firm border was established around it, establishing its area as approximately 1500 square kilometres, with the Randstad approximately 6000 square kilometres (Kühn, 2003). 70 to 80 per cent of the Green Heart is in agricultural use. The aims of the policy have been to keep the area open 'for the sake of urban dwellers and so as to form a green counterweight to the urban development around it' (van der Valk and Faludi, 1997, p. 62) and to optimise benefits for both urban and rural residents.

The planning profession has been central to promoting the Green Heart, as in the UK, and van der Valk and Faludi argue that it has become part of planning 'doctrine' in the Netherlands in much the same way as the green belt in the UK. We return to this notion in Chapter 7. But there have been criticisms of the Green Heart, perhaps on account of a contradiction in its aims identified by Kühn (2003, p. 25): it is intended to be both 'a separator of urban and rural areas and to be an integrator towards the Regional City'. To that end, transport infrastructure cuts through the Green Heart in several places, and at various points in the past it has seen higher rates of urbanisation than the Randstad (Van Eeten and Roe, 2000). Van Eeten and Roe also note that the Green Heart has a lower proportion of nature and recreational areas than the Randstad and a higher population density than the national average. Yet it persists as an important metaphor for planning in the Netherlands, and it is this metaphorical or conceptual aspect of preservation that is perhaps its strength, conveying a broad approach to urban development that is widely understood – much like the green belts in the UK.

As noted above, the Green Heart inverts the concept of the green belt. City planning in Berlin combines both an attempt to maintain a green centre and a form of green belt.

Berlin (and Germany)

The Grüne Mitte, or Green Centre, was used in the 1980s in Berlin to refer to a 'coherent complex of open spaces in the center of the city' (Lachmund, 2013, p. 166). These included the famous Tiergarten and various pieces of derelict land, some sterilised by the Berlin Wall and left undeveloped for forty to fifty years, as a consequence featuring high levels of biodiversity. Several of these sites, in both what were previously the West and East of the city, were subsequently protected as nature parks.

The designation of land as parks is also the approach used in Berlin and its broader region to protect the open spaces around the city. Berlin grew rapidly in the late nineteenth and early twentieth century, and the population increased from under 0.5 million in 1850, to 1 million in 1877, and to 2 million in 1905. Initially the city expanded in a concentric fashion, but by 1920 it was following a radial and polycentric model along or around transport links (Kühn and Gailing, 2008). Forms of green space creation or preservation were proposed in Berlin from 1840 onwards, and in 1874 the 'Countess Adelheid Dohna-Poninski called for an "outer green ring" to limit the expansion of the "compact mass of housing" in the city so that "the right of each and every inhabitant to be able to reach the open countryside within half an hour's journey from their homes could not be violated"' (cited ibid., p. 188). Competing plans for Berlin produced in 1909 included a green belt and green wedges, and competition between those proposing concentric and radial models of growth continued throughout the twentieth century (Kühn and Gailing, 2008). Of course, the damage to the city as a result of the Second World War and the subsequent division between East and West resulted in an unusual built form. After the war, as West Berlin was constrained by the Berlin Wall, it became a compact city by default. East Berlin also followed a model of development aimed at a 'compact, socialist city' (ibid., p. 187), with green wedges being planned into the city centre. This has resulted in a very high density in the urban core of the Berlin-Brandenburg region of 3,909 people per square kilometre, while that of the surrounding countryside is 175 per square kilometre (Kühn, 2003).

After 1989 and the fall of the Wall, there were calls to preserve the green space around West Berlin, as a counterpoint to the high urban densities within the city, via a green belt. Eight regional parks were proposed in 1993 – these were to be multifunctional, countryside and recreational spaces and did not have statutory protection (Kühn and Gailing, 2008). These eight parks form a ring of about 15 kilometres in radius and an area of 2500 square kilometres. They are separated only by the pre-existing settlement axes that have remained focused against any further development. Kühn and Gailing argue that this physical structure reflects the heterogeneity of the landscape around Berlin better than a monostructural green belt.

Planning in Germany is devolved to state, regional and local governments, so different regions have different approaches to green belts. 60 per cent of them have taken a regional approach to green belt planning, perhaps because 'greenbelts are an instrument of urban containment at *regional* scales' (Siedentop *et al.*,

2016, p. 81). The study by Siedentop and his colleagues of green belts around Düsseldorf, Hanover, Mittelhessen and Stuttgart found that they had succeeded in protecting environmental and recreational assets from development but less at limiting suburbanisation and urban sprawl. The authors concluded that the latter was because German green belts are, in their terms, 'rather generous' (ibid., p. 80), with space left for urban expansion within them – in contrast to the UK.

Milan, Italy

Milan, the largest metropolitan area in Italy, has been subject to a number of eras of industrial growth and decline. This has impacted on its spatial form and has led to the city and regional governments taking steps to control the 'sprawling shape' of its expansion (Senes *et al.*, 2008, p. 208). From the 1960s onwards Milan had looked to manage its landscape in conjunction with other local municipalities using structures such as the Milanese Inter-Municipal Plan, or *Piano intercomunale milanese*. Investment in green belt protection was extended through such policy structures, and further legislation was approved in 1983 and 1990 by the regional Lombardy government to protect the city's green network (Sanesi *et al.*, 2007).

In 2000 the Directorate General for Agriculture commenced on a project called the "Ten large forests for the plains" with the aim of increasing the area of forest in and around Italian cities, nominally to develop green belt areas. Milan identified ten areas for the project and in 2006 created an initiative called "10,000 hectares of forests and green systems"; this has subsequently been developed further into a plan for strategic development in the area. These two programmes provided the city government of Milan, and the wider Lombardy region, with a regional-scale project to help deliver green belt resources within a wider green network. This process had a number of aims, including the avoidance of coalescence of urban areas, the protection of ecological resources, and the promotion of a regional understanding of landscape value.

At a city scale, in 1990 the creation of the Milanese general municipal plan, *Piano regolatore generale comunale*, helped to establish the Metropolitan Belt Regional Park, also known as the South Milan Agricultural Park (Parco Sud), to protect the integrity of urban-fringe areas as a coherent green belt (Senes *et al.*, 2008). In 2002 the level of protection for the area was extended through a framework accord from the Ministry of the Environment and the Treasury for the economic programmes of the region of Lombardy, the province of Milan, and the municipality of Milan. Green belts link the Parco Sud with other landscape resources, including "metro forests", which are a key element of this process. Parco Sud is 46,300 hectares in area, has a width of approximately 15 kilometres, and is part of the 86,600 hectare metropolitan network of green infrastructure or green belt. The park is one of a series of large sites that are used as transitional areas to link the urban areas of Milan with the wider countryside, and, although larger, it acts as a connective and functional landscape element (Sanesi *et al.*, 2007). Given its size, it also requires cross-boundary commitments to management.

This is achieved through the broad Metropolitan Milano project, which coordinates the development and management objectives of different local municipalities under the wider green network of Milan (Mell, 2016).

Milan continues to invest in green infrastructure through the connection of landscape features within the green network. Urban forests, as well as linking people to parks and larger areas of the green belt, form a principal part of this process, as Milan's various levels of government attempt to retain the ecological functionality of the city and region's resource base (Spanò *et al.*, 2015). There is also an ongoing dialogue between municipalities around Milan and the city to discuss how to ensure sufficient funding and appropriate management to maintain the green belt in the long term. Adding further weight to the use of metro forests and green belt designations has been the *BioMilano* project. This project aims to redevelop brownfield and underused sites before exploiting any green space to ensure that the green network of Milan is maintained. Milan also has a history of utilising forests and woodlands as a key form of greening, as highlighted by the *Metrobosco* project or urban forestry and the *Bosco verticale* housing development (Mell, 2016). The BioMilano project is political in nature in that it has support from a number of different bodies, who view the retention of the agricultural productivity of the green belt, as well as its socio-cultural values, as essential.

Ceinture verte: Paris, France

The development of the green belt in Paris illustrates some of the difficulties in marrying disparate municipal agendas, especially where there is a perceived need to meet urban growth objectives. As the largest metropolitan area in France, Paris is potentially subject to greater development pressures than other regions. As a consequence, discussions about the green belt are framed by expansionist and protectionist narratives.

The notion of protecting a green belt around Paris was proposed as early as the 1840s, when there were calls for investment in green space on the periphery of the city. These calls were linked to the expanding urban footprint of Paris, as well as to military and political change in the city government. More recently, the creation of a green belt has been implemented in two stages: the development of a ring of green space around the greater Paris region (*Ceinture verte* – the green belt) and a more discreet network of green infrastructure within Paris itself (the urban core), the latter making use of the city's rivers and waterways as blue infrastructure. Understanding the spatial footprint of the city's green space has allowed researchers to explore the value of the environment as it represents the "physical landscape" context of both Paris and the Île-de-France region within which it sits. Such a spatial interpretation has made it possible for decision-makers, academics and the public to reflect on the green belt in terms of offering social, economic and ecological benefits to local and wider areas (Laruelle and Leganne, 2008).

In 1976 the Regional Natural Parks were created as a mechanism to ensure that the region's landscape remained rural in nature. The development of these parks was also viewed as helping to retain the agricultural function of the Île-de-France, both economically and ecologically, without sacrificing its environmental resource base to development. Furthermore, it has been reported that the rationale behind the spatial understanding of the green belt was to minimise the infill of the gaps between existing green wedges. However, there was, and continues to be, scope to develop in green belt areas if the investment is deemed complementary to existing infrastructure (APUR, 2013).

While the idea of the green belt has been discussed for a number of decades in Paris, it was legislated for only in 1994, when a green belt of 300,000 hectares was formally created to form a green space network linking agricultural, 'natural', and woodland areas (Laruelle and Leganne, 2008). From 2000 onwards there have been ongoing discussions as to whether the green belt should be expanded into a wider regional-scale belt (*Plan vert régional*). This highlights one of the unusual aspects of the Paris green belt: its expansive nature. Three distinct levels – regional, intermunicipal and local – can be identified, all of which have an influence on how the green belt can and does function and is managed. The central urban *trame verte* (green grid) covers 5 per cent of the region's total surface area (10 per cent of the area's green space), the *ceinture verte*/green belt covers 22 per cent of the area, while the *couronne rurale* (rural outer ring) covers 73 per cent of the region's total surface area (90 per cent of its green infrastructure) (ibid.).

Over time, though, as growth has occurred, there has been a changing emphasis and value placed upon the green belt, putting it under pressure to deliver urban development. There are also ongoing discussions concerning the nature of green belt management in the Île-de-France, as some consider that the designation is somewhat discretionary and flexible in terms of what development may or may not be permitted (APUR, 2013). The Atelier parisen d'urbanisme has been a key actor developing green belt discussions in Paris. Using the "Paris's Green Belt in the 21st Century" proposals, it aims to increase the proportion of nature in the green belt and the percentage of accessible green spaces in the wider Paris region, thus improving the connectivity between urban and rural areas. However, because of the range of landowners and land uses, it has proved difficult for the area's planning authorities to manage the *ceinture verte* as a coherent entity (Laruelle and Leganne, 2008).

North America

The USA

One-quarter of metropolitan areas in the USA now use some form of containment policy to manage, limit or control urban growth (Dawkins and Nelson, 2002), usually on the basis of the 1928 Standard Zoning Enabling Act, which gives local authorities the power to regulate land use (Carruthers, 2002). With a few exceptions, these policies are not typically green belts *per se*. More common are urban growth boundaries (UGBs), effectively lines on the map that dictate how far an urban area can expand, akin to the inner boundary of a green belt. The first UGB was established in Lexington, Kentucky, in 1958; by 1999, 100+ cities and counties had UGBs, and three states (Oregon, Tennessee and Washington) had mandated them (Jun, 2004).

Three specific examples are discussed below, but Carruthers (2002) undertook a broad literature review of a range of 'regulatory growth management programs' in the USA. He concluded that the level at which such programs originate makes a difference as to their effects, with extra-local initiatives, whether at the state or regional level, being more effective. Local growth management could be parochial, in the absence of regional planning which considered how the wider area was developing, if it was simply aimed at limiting the amount of growth in a particular location; this growth could simply be diverted elsewhere, in some cases promoting urban sprawl.

The first case from the USA we wish to discuss is that of Portland and its home state, Oregon.

Portland (and Oregon)

The Oregon system involves a state commission, introduced in 1973, which coordinates and approves city, county and special district plans (Knaap, 1985) through Senate Bill 100 (Dawkins and Nelson, 2002). The principle of urban growth boundaries was introduced at the same time. When a UGB has been designated, only land within it can be converted to urban use before a specified date. Land outside it is preserved until after that date (Knaap, 1985). The 1973 Bill *requires* local governments in Oregon to adopt UGBs, but policy-makers were wary of the political consequences of restricting growth across the state, so included policies to ensure that UGBs did not constrain land supply for housing and the provision of jobs (Dawkins and Nelson, 2002).

The Portland UGB, covering 227,410 acres (91,874 hectares), was adopted in 1979 (Jun, 2004, 2006) and, in an example of avoiding constraining growth, has expanded virtually every year (Metro, 2015b). By 2014 this had resulted in a 14 per cent increase in its area (Metro, 2015a), to 258,796 acres (104,554 hectares). The evidence suggests that it has been effective in managing urban growth: the urbanised population has risen by 54 per cent while, as noted above, the area of the UGB has expanded by only 14 per cent. The density of the built form has increased, with the proportion of developed land within the UGB going up by 36 per cent between 1980 and 2000 and the population density rising by 13.6 per cent (Jun, 2004). As noted in Chapter 4, various studies have explored the effect of the Portland UGB on house prices, but with no clear result.

The Portland UGB is managed by a directly elected regional government (the only such government in the USA), so cutting across municipal boundaries and having an effect on the regional land market. There remain, however, cross-border issues, as part of the metropolitan area of Portland is in Clark County, in the adjacent state of Washington. Although Oregon and Washington are the only pair of adjacent states to adopt state-wide growth management policies, Washington introduced growth controls some time after Oregon. Bae investigated whether there was overspill into Clark County and concluded that it was 'not an accident' (2004, p. 110) that the latter was the fastest growing county in the state of Washington, as well as among the four counties that comprise Portland.

Boulder, Colorado

There is a long history of land preservation in Boulder, Colorado, as the city bought its first piece of parkland for preservation in 1898 (Hickcox, 2009).The population doubled to 40,000 between 1950 and 1960 (Belford, 2013) and again to 80,000 by 1970 (Hickcox, 2009). In 1967 this led Boulder to establish, after a public vote, a fund to buy and manage green belts, financed by a 0.4 per cent city sales tax. In 1978 this generated $1 million per year in revenue and allowed the city to buy a total of 8000 acres (3232) hectares of land (Correll *et al.*, 1978). A further $200 million was subsequently spent to acquire a total of more than 45,000 acres (18,180 hectares) (Hickcox, 2009).

This has led to particular patterns of growth, and little housing has spread onto the mountainous areas that surround the city. Boulder also has an unusual demographic profile, with a household income twice the national average in 2000 and a 'relatively small number of racial or ethnic minorities' (Hickcox, 2009, p. 237). While some point to the latter as being an unintended consequence of the green belt and other protection policies, Hickcox argues that, since the early twentieth century, the city has deliberately excluded working-class and non-white residents, using the green belt to do so. We do not wish to comment either way on this argument: whichever way the causal link flows, the Boulder case illustrates that green belts and other growth management policies do have the potential to be regressive, as some have found of similar policies in California.

California

There is no requirement from the state in California to implement UGBs or any other form of urban containment, but many local authorities have done so, often 'for the explicit *purpose* of restricting the local supply of housing' (Dawkins and Nelson, 2002, p. 8, emphasis in original). Many of these policies were introduced in the late 1970s in response to *Proposition 13*, which set tight limits on property taxation in California, resulting in local governments having a growing population but a shrinking tax base (Simmie, 1993). Such policies have been supported politically by local residents to 'reduce the numbers of new houses built in their areas' (ibid., p. 55). Studies by Simmie and by Dawkins and Nelson found that, if such policies were successfully implemented, the price of housing tends to go up. In some places, for example Sacramento, the UGB is amended as necessary, having no effect on house prices. Simmie found that the causal link between growth management and house prices was indirect and as a result of oligopolies of large house-building firms dominating the market, in part because of growth management generating barriers to entry for other, smaller, firms. His conclusion was that increases in house prices subsequently resulted in residential segregation, with poorer households (who, statistically, are more likely to be non-white) forced to move further from their places of work in order to find affordable accommodation.

Canada

Toronto/Horseshoe Bay green belt

The Greater Golden Horseshoe in Ontario is potentially the world's largest green belt, as it has permanent legislative protection to an area of over 1.8 million acres (7300 square kilometres) of land surrounding the city of Toronto (Carter-Whitney and Esakin, 2010). The designation was proposed in 2003 by the then premier of Ontario, Dalton McGuinty, and in policy in the 2004 Green Belt Protection Act. It was subsequently passed into law by the Green Belt Act (2005). The aim of the green belt was to promote a prosperous and sustainable Ontario that would be able to supply some of its own food resources. The legislation was also intended to protect agricultural land, heritage sites, and sensitive ecology and hydrology (including the Niagara Escarpment and the Oak Ridges Moraine, the latter a large area of ecological and geological interest), to avoid coalescence and to retain the region's urban areas within a growth boundary (Cadieux *et al.*, 2013; Pond, 2009). The green belt further provides socio-economic amenities for the region (Macdonald and Keil, 2012).

A strong "Friends of the Green Belt Foundation", a non-profit organisation, was established in 2005 to promote and lobby for the wider benefits of the green belt. A Green Belt Council was also formed to oversee the strategic development of the project, coordinated by the Ministry of Municipal Affairs and Housing, and has garnered high-level support from both landowners and agricultural businesses (Pond, 2009).

The ten-year anniversary of the green belt passing into legislation fell on 28 February 2015. This led to a review of its utility and functionality being undertaken to coincide with a review of the Growth Plan for the Greater Golden Horseshoe, the Niagara Escarpment Plan and the Oak Ridges Moraine Conservation Plan. Not all representations to the process were positive. Many other commentators, however, have noted the added ecological and social value that can be attributed to the green belt designation. Reeves (2015), for example, suggested that one of the key impacts of the green belt was to slow habitat fragmentation to limit ecological isolation. It also protects wetlands, river valleys, lakes and forest, as well as seventy-eight threatened or at risk species which support the key landscape features of the Ontario region (Taylor *et al.*, 1995). Moreover, Reeves (2015) argued that the green belt provides up to CAN$9 billion in direct and indirect benefits to businesses and an additional $2.8 billion from the 161,000 Ontarians employed through green belt-related activities. Almack and Wilson (2010) went further, stating that the natural capital value of the green belt of $3487 per hectare was important in delivering a functional and attractive landscape with wider benefits to the public, politicians and business.

However, a counter-argument has been made against the green belt, as some commenters, including Curtis (2014) and Cox (2004), suggest that it has a negative impact on the affordability of housing and the development of transport infrastructure and places restrictions on commercial development. Thus the growth needs of the Greater Golden Horseshoe area are at odds with the preservationist views of the green belt. Mausberg (2014), though, argued that the green belt has 90 per cent positive feedback in government polls, but he did concede that the cost of mortgages (due to low interest rates) and people's aspiration to own suburban single family homes has led to calls for release and conversion of green belt land.

Although there have been appeals to convert sections of the green belt for urban development, Keesmaat and Burda (2015) noted that, in each of the two ten-year periods 1991–2001 and 2001–11, the population of Toronto had grown by 1 million people. In the first period, this led to a 26 per cent increase in land use, while in the second period the increase was only 10 per cent. They go on to state that the main criticisms of the green belt have been economic in nature, focusing most frequently on house prices. However, such comments often overlook the socio-ecological benefits of the green belt by claiming a lack of social equity in use and exposure. The authors conclude that building in the green belt is not the only answer to housing or infrastructure needs. Moreover, Ballingall (2015) argues that there are also demands for an expansion of the green belt with a further 1.5 million acres of 'bluebelts', to include the Oro Moraine and Humber River Headwaters, that would benefit the area ecologically and socio-economically. This would make the green belt spatially bigger – to a catchment level – but would need to take into consideration the growing population and demands of these people.

Conclusions

As we identified in the opening section of this chapter, the context within which green belts have been used is different in every circumstance. Therefore what each of the case studies presented here has done is to look at the manner in which green belts are presented in policy and development discussions and identify both where good practice has occurred and where conflicts have arisen in their use. Three common themes emerge from this review.

The first is that *scale matters*. Some green belts, such as the Golden Horseshoe, are landscape-scale designations, while others, for example that around Berlin, are smaller urban-orientated sites. The variability in size allows local government and their planners to work with the landscape to identify the most appropriate form of investment and protection. Moreover, smaller designations are less complicated because they require more discreet buy-in and support from stakeholders (Watanabe *et al.*, 2008). It may also be easier to control the management of a smaller green belt. As the examples of Toronto and Milan highlight, though, those green belts that work at a landscape scale can generate greater visibility, which in turn can lead to increased political, social and economic support (Cox, 2004; Mell, 2016). Further, there is evidence that policies at the extra-local level can be more effective because the various urban areas do not work in isolation, so overspill effects are minimised (Carruthers, 2002).

Secondly, policies focused exclusively on limiting urban growth, and not on the positive uses of green spaces, work less well. Because of the variability in the appearance of the green belt and the functions it delivers, there is a corresponding flexibility in how they are managed. This is evident in the green belts of Paris and Milan. Where agricultural land is present, conflicts may arise between the need to maintain landscape productively, to promote access to nature, and to provide land for development. Milan has managed this process through the planting of new forest areas (Mell, 2016). Similarly, in the Netherlands the protection of the Green Heart has meant that calls for widespread conversion of natural or agricultural landscapes to meet housing and infrastructure needs have been resisted (Van Eeten and Roe, 2000). Other places, including New Delhi, have struggled to ensure that existing landscape functionality is maintained in the face of ongoing discussions of development and growth (Kumar, 2015; Narain, 2009). In Ahmedabad, the rescinding of green belt policy due to growth pressures has led to criticism of the city's Urban Development Authority and Metropolitan Council (Manthur, 2012). Moreover, the decisions of local government in Seoul, Tokyo and Sydney have seen the proportion of land designated as green belt shrink as a result of a perceived need to meet infrastructure needs (Bae and Jun, 2003; Evans and Freestone, 2010; Watanabe *et al.*, 2008). However, if green belt policy is aligned with development and conservation activities, such as in Boulder and parts of California, we see a more reflective form of landscape management. This, though is difficult in many locations, including Beijing and Shanghai, where the rapidity of expansion is such that it becomes politically and economic challenging to ensure green belt designations are maintained (Yang and Jinxing, 2007; Zhao, 2011).

Thirdly, the aim/purpose of green belts affects how they are implemented and subsequently how their success or otherwise is evaluated. In each of the case studies, alternative views of the value and need for green belt resources exist. For example, in Toronto these relate to real-estate prices (Cox, 2004), in Berlin to the redevelopment of derelict or undervalued spaces (Lachmund, 2013), and in the USA to the influence that UGBs have on expansion (Carruthers, 2002; Knapp, 1985). We can identify discussions of the variability of understanding of green belts around the world comparable to those concerning the UK in Chapter 3. However, while this provides a certain amount of flexibility for planners to designate and implement green belt policy, it can also restrict investment owing to a lack of clarity in the perceived outcomes of development (Amati and Taylor, 2010; Amati, 2008).

So how much can we learn from the case studies that might be applicable to the UK experience? The extent to which each one is similar to, or different from, what happens in the UK differs. In some cases, such as Melbourne and the Netherlands, explicit direct inspiration was taken from the UK in founding the green belt (or the Green Heart). In others, there are similarities in how the green belt has been developed which make comparison easier – in Paris, for example, the long gestation of green belt since the mid-nineteenth century suggests a similar history of concern about the impacts of industrialisation and urbanisation. In China, though urbanisation has arrived much more recently, scholars have noted the difficulty in adapting a state-led planning system to the new reality of private-sector-led development (Yang and Jinxing, 2007) – not dissimilar to the criticisms discussed in Chapter 2 of the failure to adapt the plans of Abercrombie and others to a different context. In the Netherlands, the Green Heart has become part of planning doctrine (van der Valk and Faludi, 1997), and thus is cemented into the philosophy of the planning profession and the national culture alike, making attempts at reform difficult; the comparison with the UK, as discussed in the conclusion to Chapter 2, is striking.

Notwithstanding the difficulties in enacting reform, the following chapter asks what alternatives there are for planners in the UK to using green belts as a form of landscape and urban planning.

References

Adhvaryu, B. (2011) The Ahmedabad urban development plan-making process: a critical review, *Planning Practice and Research*, 26(2): 229–250.

Ahmedabad Urban Development Authority (2013) *Draft Comprehensive Development Plan 2021* (2nd revision, Ahmedabad: AUDA).

Almack, K., and Wilson, S. (2010) *Ontario's Wealth, Canada's Future: The Economic Value of the Greenbelt Plan in Toronto, Canada*, www.TEEBweb.org.

Amati, M. (ed.) (2008) *Urban Green Belts in the Twenty-First Century* (Aldershot: Ashgate).

Amati, M., and Taylor, L. (2010) From green belts to green infrastructure, *Planning Practice and Research*, 25(2): 143–55.

APUR (2013) *Le ceinture verte de Paris au XXIe siècle: hier, aujourd'hui, demain* (Paris: Atelier parisien d'urbanisme); www.apur.org/sites/default/files/documents/ceinture_verte_paris_chapitre1_0.pdf.

Bae, C.-H. C. (1998) Korea's greenbelts: impacts and options for change, *Pacific Rim Law and Policy Journal*, 7(3): 479–502.

Bae, C.-H. C. (2004) Cross-border impacts of a growth management regime: Portland, Oregon, and Clark County, Washington, in A. Sorensen, P. J. Marcotullio and J. Grant (eds), *Towards Sustainable Cities: East Asian, North American and European Perspectives on Managing Urban Regions* (Aldershot, Ashgate), pp. 95–111.

Bae, C.-H. C., and Jun, M.-J. (2003) Counterfactual planning: what if there had been no greenbelt in Seoul?, *Journal of Planning Education and Research*, 22(4): 374–83.

Ballingall, A. (2015) Environmental groups hope to nearly double size of the greenbelt, *Toronto Star*, 5 November, www.thestar.com/news/2015/11/05/environmental-groups-hope-to-nearly-double-size-of-the-greenbelt.html.

Bangalore Development Authority (2005) *Draft Master Plan – 2015: An Integrated Planning Approach ... towards a Vibrant International City*, www.bdabangalore.org/brochure.pdf.

Belford, S. (2013) Open space as a homegrown ideal: Boulder, Colorado's 1967 greenbelt amendment, https://centerwest.org/wp-content/uploads/2013/05/FirstUNDER ACADEMICNONFICTbelfordOpenSpace.pdf.

Bengston, D. N., and Youn, Y. C. (2006) Urban containment policies and the protection of natural areas: the case of Seoul's greenbelt, *Ecology and Society*, 11(1), www.ecologyandsociety.org/vol11/iss1/art3/.

Buxton, M. (2014) The expanding urban fringe: impacts on peri-urban areas, Melbourne, Australia, in B. Maheshwari, R. Purohit, H. Malano, V. P. Singh, and P. Amerasinghe (eds), *The Security of Water, Food, Energy and Liveability of Cities* (Dordrecht: Springer), pp. 55–70.

Buxton, M., and Goodman, R. (2008) Protecting Melbourne's green wedges – fate of a public policy, in M. Amati (ed.), *Urban Green Belts in the Twenty-First Century* (Aldershot: Ashgate), pp. 61–82.

Cadieux, K. V., Taylor, L. E., and Bunce, M. F. (2013) Landscape ideology in the Greater Golden Horseshoe greenbelt plan: negotiating material landscapes and abstract ideals in the city's countryside, *Journal of Rural Studies*, 32: 307–19.

Carruthers, J. I. (2002) Evaluating the effectiveness of regulatory growth management programs: an analytic framework, *Journal for Planning Education and Research*, 21(4): 391–405.

Carter-Whitney, M., and Esakin, T. (2010) *Ontario's Greenbelt in an International Context* (Toronto: Canadian Institute for Environmental Law and Policy).

Chatterji, T. (2013) The micro-politics of urban transformation in the context of globalisation: a case study of Gurgaon, India, *South Asia: Journal of South Asian Studies*, 36(2): 273–87.

Chaudhry, P., Bagra, K., and Singh, B. (2011) Urban greenery status of some Indian cities : a short communication, *International Journal of Environmental Science and Development*, 2(2): 98–101.

Choe, S.-C. (2004) Reform of planning controls for an urban–rural continuum in Korea, in A. Sorensen, P. J. Marcotullio and J. Grant (eds), *Towards Sustainable Cities: East Asian, North American and European Perspectives on Managing Urban Regions* (Aldershot: Ashgate), pp. 253–66.

Correll, M. R., Lillydahl, J. H., and Singell, L. D. (1978) The effects of greenbelts on residential property values: some findings on the political economy of open space, *Land Economics*, 54(2): 207–17.

Cox, W. (2004) *Myths about Urban Growth and the Toronto "Greenbelt"*, www. demographia.com/db-torgreenbelt.pdf.

Curtis, T. (2014) Has Toronto's greenbelt done more harm than good?, *The Globe and Mail*, 21 November, www.theglobeandmail.com/opinion/has-torontos-greenbelt-done-more-harm-than-good/article21689009/.

Dash, D. (2014). UP, Haryana told to increase forest cover in NCR towns, *Times of India*, 13 April, http://timesofindia.indiatimes.com/city/delhi/UP-Haryana-told-to-increase-forest-cover-in-NCR-towns/articleshow/33672125.cms.

Dawkins, C. J., and Nelson, A. C. (2002) Urban containment policies and housing prices: an international comparison with implications for future research, *Land Use Policy*, 19(1): 1–12.

Delhi Development Authority (2007) *Master Plan for Delhi – 2021*, https://dda.org.in/ddanew/pdf/Planning/reprint%20mpd2021.pdf.

Evans, C., and Freestone, R. (2010) From green belt to green web: regional open space planning in Sydney, 1948–1963, *Planning Practice and Research*, 25(2): 223–40.

Ghaziabad Development Authority (2005) *Ghaziabad Master Plan – 2021*, http://assetyogi.com/resources/master-plans/up/ghaziabad-master-plan-2021/.

Government of India, Ministry of Environment Forest and Climate Change (1980) *Forest (Conservation) Act, 1980 with Amendments Made in 1988*, www.moef.nic.in/sites/default/files/Forest.pdf.

Hickcox, A. (2009) Green belt, white city: race and the natural landscape in Boulder, Colorado, *Discourse*, 29(2): 236–59.

India Today (2010) Pay compensation for commerical use of green belt land, SC, *India Today*, 25 March, http://indiatoday.intoday.in/story/Pay+compensation+for+commerical+use+of+green+belt+land,+SC/1/89886.html.

Jain, A. (2014) NGT to ascertain concretisation of "green belts", *The Hindu*, 30 September, www.thehindu.com/news/cities/Delhi/ngt-to-ascertain-concretisation-of-green-belts/article6460486.ece.

Jayaseelan, N. (2015) 6% forest cover insufficient for a 62% urbanised NCR, *Hindustan Times*, 30 June, www.hindustantimes.com/ht-view/6-forest-cover-insufficient-for-a-62-urbanised-ncr/story-y0XyVtHQ5Y0YEPmtfa5C0O.html.

Jongman, R., and Pungetti, G. (eds) (2004) *Ecological Networks and Greenways: Concept, Design and Implementation* (Cambridge: Cambridge University Press).

Jun, M.-J. (2004) The effects of Portland's urban growth boundary on urban development patterns and commuting, *Urban Studies*, 41(7): 1333–48.

Jun, M.-J. (2006) The effects of Portland's urban growth boundary on housing prices, *Journal of the American Planning Association*, 72(2): 239–43.

Keesmaat, J., and Burda, C. (2015) Greenbelt makes GTA more, not less, livable, *Toronto Star*, 17 March, www.thestar.com/opinion/commentary/2015/03/17/greenbelt-makes-gta-more-not-less-livable.html.

Kim, C. H., and Kim, K. H. (2000) The political economy of Korean government policies on real estate, *Urban Studies*, 37(7): 1157–69.

Kim, J., and Kim, T.-K. (2008) Issues with green belt reform in the Seoul metropolitan area, in M. Amati (ed.), *Urban Green Belts in the Twenty-First Century* (Aldershot: Ashgate), pp. 37–57.

Knaap, G. J. (1985) The price effects of urban growth boundaries in metropolitan Portland, Oregon, *Land Economics*, 61(1): 26–35.

Kühn, M. (2003) Greenbelt and Green Heart: separating and integrating landscapes in European city regions, *Landscape and Urban Planning*, 64(1–2): 19–27.

Kühn, M., and Gailing, L. (2008) From green belts to regional parks: history and challenges of suburban landscape planning in Berlin, in M. Amati (ed.), *Urban Green Belts in the Twenty-First Century* (Aldershot: Ashgate), pp. 185–202.

Kumar, S. (2015) Threat to green belt in South Delhi: NGT notice to Delhi government, *Economic Times*, 31 August, http://articles.economictimes.indiatimes.com/2015-08-31/news/66070476_1_green-belt-boundary-wall-gurudwara.

Lachmund, J. (2013) *Greening Berlin: The Co-Production of Science, Politics, and Urban Nature* (Cambridge, MA, and London: MIT Press).

Laruelle, N., and Leganne, C. (2008) The Paris–Île-de-France ceinture verte, in M. Amati (ed.), *Urban Green Belts in the Twenty-First Century* (Aldershot: Ashgate), pp. 227–241.

Lee, C. M. (1999) An intertemporal efficiency test of a greenbelt: assessing the economic impacts of Seoul's greenbelt, *Journal of Planning Education and Research*, 19(1): 41–52.

Lee, C. M., and Linneman, P. (1998) Dynamics of the greenbelt amenity effect on the land market: the case of Seoul's greenbelt, *Real Estate Economics*, 26(1): 107–29.

Macdonald, S., and Keil, R. (2012) The Ontario greenbelt: shifting the scales of the sustainability fix?, *Professional Geographer*, 64(1): 125–45.

Mathur, N. (2012) On the Sabarmati riverfront: urban planning as totalitarian government in Ahmedabad, *Economic and Political Weekly*, 47(47–8), 64–75.

Mausberg, B. (2014). Don't blame Ontario's greenbelt for GTA's rising house prices, *Toronto Star*, 8 January, www.thestar.com/opinion/commentary/2014/01/08/dont_blame_ontarios_greenbelt_for_gtas_rising_house_prices.html.

Mees, P. (2003) Paterson's curse: the attempt to revive metropolitan planning in Melbourne, *Urban Policy and Research*, 21(3): 287–99.

Mell, I. C. (2016) *Global Green Infrastructure: Lessons for Successful Policy-Making, Investment and Management* (Abingdon: Routledge).

Metro (2015a) Portland Metropolitan Urban growth boundary expansion history, www. oregonmetro.gov/sites/default/files/UGB_History.pdf.

Metro (2015b) Urban growth boundary, www.oregonmetro.gov/urban-growth-boundary.

Nagendra, H., and Gopal, D. (2010) Street trees in Bangalore: density, diversity, composition and distribution, *Urban Forestry & Urban Greening*, 9(2): 129–37.

Nagendra, H., Nagendran, S., Paul, S., and Pareeth, S. (2012) Graying, greening and fragmentation in the rapidly expanding Indian city of Bangalore, *Landscape and Urban Planning*, 105(4): 400–06.

Narain, V. (2009) Growing city, shrinking hinterland: land acquisition, transition and conflict in peri-urban Gurgaon, India, *Environment and Urbanization*, 21(2): 501–12.

Pond, D. (2009) Ontario's greenbelt: growth management, farmland protection, and regime change in southern Ontario, *Canadian Public Policy*, 35(4): 413–32.

Press Trust of India (2015) NGT prohibits tree felling in a south Delhi green belt, *Business Standard*, 2 December, www.business-standard.com/article/pti-stories/ngt-prohibits-tree-felling-in-a-south-delhi-green-belt-115120200321_1.html.

Reeves, A. (2015).What's next for Ontario's greenbelt?, *Torontist*, 21 October, http://torontoist.com/2015/10/whats-next-for-ontarios-greenbelt/.

Ren, W., Zhong, Y., Meligrana, J., Anderson, B., Watt, W. E., Chen, J., and Leung, H. L. (2003) Urbanization, land use, and water quality in Shanghai, 1947–1996, *Environment International*, 29(5): 649–59.

Rushman, J. (1969) Melbourne and Sydney: problems of continuing urban growth, *Town Planning Review*, 40(3): 263–82.

Sanesi, G., Lafortezza, R., Marziliano, P., Ragazzi, A., and Mariani, L. (2007) Assessing the current status of urban forest resources in the context of Parco Nord, Milan, Italy, *Landscape and Ecological Engineering*, 3(2): 187–98.

Senes, G., Toccalini, A., Ferrario, P., Lafortezza, R., and Dal Sasso, P. (2008) Controlling urban expansion in Italy with green belts, in M. Amati (ed.), *Urban Green Belts in the Twenty-First Century* (Aldershot: Ashgate), pp. 203–226.

Siedentop, S., Fina, S., and Krehl, A. (2016) Greenbelts in Germany's regional plans: an effective growth management policy?, *Landscape and Urban Planning*, 145(1): 71–82.

Simmie, J. M. (1993) *Planning at the Crossroads* (London: UCL Press).

Spanò, M., DeBellis, Y., Sanesi, G., and Lafortezza, R. (2015) *Milan, Italy: Case Study City Portrait*, http://greensurge.eu/products/case-studies/Case_Study_Portrait_Milan.pdf.

State of Victoria, Department of Environment, Land, Water and Planning (2008) *Melbourne 2030: A Planning Update – Melbourne @ 5 Million*, www.dtpli.vic.gov.au/planning/plans-and-policies/planning-for-melbourne/melbournes-strategic-planning-history/melbourne-2030-a-planning-update-melbourne-@-5-million.

State of Victoria, Department of Infrastructure (2002) *Melbourne 2030: Planning for Sustainable Growth*, www.dtpli.vic.gov.au/planning/plans-and-policies/planning-for-melbourne/melbournes-strategic-planning-history/melbourne-2030-planning-for-sustainable-growth.

Sudha, P., and Ravindranath, N. (2000) A study of Bangalore urban forest, *Landscape and Urban Planning*, 47(1–2): 47–63.

Taylor, J., Paine, C., and FitzGibbon, J. (1995) From greenbelt to greenways: four Canadian case studies, *Landscape and Urban Planning*, 33(1–3): 47–64.

Tian, L., and Ma, W. (2009) Government intervention in city development of China: a tool of land supply, *Land Use Policy*, 26(3): 599–609.

Times of India (2015) NGT notice on Ghaziabad green belt squatters, *Times of India*, 24 September, http://timesofindia.indiatimes.com/city/noida/NGT-notice-on-Ghaziabad-green-belt-squatters/articleshow/49082455.cms.

United Nations Department of Economic and Social Affairs (2014) *World Urbanization Prospects* (New York: United Nations).

van der Valk, A., and Faludi, A. (1997) The Green Heart and the dynamics of doctrine, *Netherlands Journal of Housing and the Built Environment*, 12(1): 57–75.

Van Eeten, M., and Roe, E. (2000) When fiction conveys truth and authority: the Netherlands Green Heart planning controversy, *Journal of the American Planning Association*, 66(1): 58–67.

Wang, H.-B., Li, H., Ming, H.-B., Hu, Y.-H., Chen, J.-K., and Zhao, B. (2014) Past land use decisions and socioeconomic factors influence urban greenbelt development: a case study of Shanghai, China, *Landscape Ecology*, 29(10): 1759–70.

Watanabe, T., Amati, M., Endo, K., and Yokohari, M. (2008) The abandonment of Toyko's green belt and the search for a new discourse of preservation in Tokyo's suburbs, in M. Amati (ed.), *Urban Green Belts in the Twenty-First Century* (Aldershot: Ashgate), pp. 21–37.

Yang, J., and Jinxing, Z. (2007) The failure and success of greenbelt program in Beijing, *Urban Forestry & Urban Greening*, 6(1): 287–96.

Zhang, T. (2000) Land market forces and government's role in sprawl: the case of China, *Cities*, 17(2): 125–35.

Zhao, P. (2011) Managing urban growth in a transforming China: evidence from Beijing, *Land Use Policy*, 28(1): 96–109.

Zhao, P., Lu, B., and De Roo, G. (2010) Performances and dilemmas of urban containment strategies in the transformation context of Beijing, *Journal of Environmental Planning and Management*, 53(2): 143–61.

6 Alternatives to green belts

Introduction

Previous chapters have discussed the history of green belts, their characteristics and their impacts. Based on these discussions, we argue that there are a series of alternatives to green-space planning which could be more appropriate forms of landscape management. This chapter looks at several approaches to green-space management in and around urban areas which are being applied successfully and could also be employed in a UK context. The aim here is to postulate whether appropriate and practicable alternatives can be identified to green belts which would retain their spatial dimension but also act more responsively to the changing needs of urban and rural communities. Each of the options proposed suggests that, if we can move beyond the current polarised discussion of green belts within the UK and elsewhere, more appropriate forms of sustainable landscape management could be implemented (Amati and Taylor, 2010).

The following analysis will make use of a series of examples to argue that there is a range of policy choices available to LPAs that might address the politically restrictive nature of green belt policy, as well as outlining some of the socio-economic and ecological benefits that they can deliver. However, because of variation in political support, existing planning structures and the nature of planning for the environment, we are not advocating that new forms of landscape management be parachuted in to take the place of green belts *per se*; alternatively, we offer a set of delivery and management options which could be used by planners, developers and landowners to maintain what is valuable about the green belt, while diversifying what may currently be mono-functional areas of land (James *et al.*, 2009).

A number of concepts will be discussed in this chapter, including *green wedges*, greenspace *networks*, so-called *green necklaces*, and strong *zoning* arrangements. This selection is not exhaustive, as additional approaches to green-space planning have been successfully implemented around the world under a 'green infrastructure' banner (Amati and Taylor, 2010; Zmelik *et al.*, 2011; Mell, 2016). Those presented are therefore our interpretation of the most common and well-researched forms of green-space planning currently being debated in the academic and practitioner literature.

The value of such a debate lies in the detail of what these approaches are attempting to do. Are they restricting growth? Are they dividing landscapes into manageable units of specific size or value? Are they exploring the role of ecological networks espoused by landscape ecologists as a mechanism to ensure value is maintained across a wider area? Or are these approaches simply a reappraisal of what green belts are charged with achieving in the UK? As reported in Chapter 5, how green belts are employed is internationally variable, which raises the question of whether the rationale for their use can be transferable between locations. Chapter 3 also drew on the conceptual differences in how green belts are viewed, which supports our discussion of the potential value of alternative green-space practices in meeting local and more strategic needs in planning (Austin, 2014; Mell, 2015a). As a consequence, there are a number of caveats in the presumption that green belts and their alternatives are geospatially and conceptually defined. Because of the variability in form that different approaches to investment in green space can take, we discuss a series of examples of how green wedges, greenways and zoning ordinances could be applied in the UK in place of the existing green belt. This is not to suggest that they should necessarily replace the current designations – rather, that they offer a more responsive set of options for LPAs, developers and local communities to engage with. The chapter asks whether it would be more viable for LPAs to utilise a range of alternative green-space practices, as they offer more flexibility to landscape and urban planning, and not rely solely on existing green belt designations.

Spatial differences between green belt and other forms of green-space planning: size, location and function

One of the key issues, as we have mentioned throughout, is the variable character of the green belt, both in the UK and internationally. It has therefore proved difficult, in our opinion, to plan legitimately for a uniform investment of approach to its management. Alternatively, we would call for a more pluralistic approach to green belt development and protection attuned to the nuances of local landscape and the socio-economic and political contexts. Such approaches are evident, and were discussed in Chapter 5, where, for example, the New Delhi green belt (The Ridge) is defined by the geography of western Delhi (Jain, 2014; Mell, 2016), and in Shanghai, where the adoption of green "loops" or "rings" to define the extent of the urban and the transition to the "rural" is a political gesture made by planners to control the zoning of the two (Zhang, 2005; Wu, 2015). Furthermore, the conceptual development of Frederick Law Olmsted's Emerald Necklace in Boston (USA) offers a more urban-centred approach to a linear belt, as it was designed to link the Charles River in downtown Boston to the then urban-fringe areas of the city. Its role as a flood mitigation system has subsequently been extended as additional climatic change, recreational and economic benefits have been attributed to the project (Ryan *et al.*, 2002; Fábos, 2004). Moreover, the green belt in the UK appears to draw heavily on notions of political control of boundaries and management of the urban–rural interface. It is also imbued with specific, and often divisive, socio-economic characteristics, which are not always taken into account (Amati, 2008). How we manage the interactions of these often conflicting variables is thus one of the most prominent issues in current green belt debates.

(a)

(b)

(c)

Figures 6.1a–c The Boston Emerald Necklace (Source: Authors' own)

In the exploration of greenways by Hellmund and Smith (2006), the complexity of aligning such variations is presented in the authors' extensive review of the terminology and types of green-space planning currently in use around the world. They note that green belts are used to 'protect natural or agricultural lands to restrict or direct metropolitan growth' (2006, pp. 2–3), and cite Boulder (Colorado) and London (UK) as examples of this in practice. However, they also discuss the role of green fingers in Buffalo (USA), green links in British Columbia (Canada), the Green Heart in the Randstad (the Netherlands) and green wedges in Melbourne (Australia) as examples of how landscape features can be used to conserve, protect and mitigate the impacts of development in and around urban–rural interfaces. What is noticeable in this discussion is the breadth of terminology used, which raises the following questions:

1   Can we define a universal form of green-space management based on comparable characteristics to green belts?
2   What evidence is available as to the use of alternative green-space planning mechanisms to achieve the socio-economic, ecological and political benefits of green belts?
3   Can we identify one or more suitable mechanisms to assist planners, developers and communities to translate the principles of green space and green belt theory into practice?

The following discussion addresses each of these questions using a series of green-space planning approaches as counterpoints to our overarching rethinking of the utility of green belts. Using examples from the categories presented by Hellmund and Smith (2006), we investigate where best practice in green-space management occurs, as well as the spatial form that it takes. This, subsequently, allows us to address the complexities of translating practice between specific geopolitical locations and assesses whether such transferability of practice is possible to meet localised green belt issues in the UK and further afield. Drawing on the wide-ranging literature, the ensuing debates make comparisons between green belts and different approaches to green infrastructure planning, examining their function, location and size. This gauges both the value people place on alternative forms of green-space planning and how the current debates surrounding green belts could be shaped by a more open-minded and inclusive approach to landscape planning.

Greenways

In comparison with the UK, green belts, as a defined form of landscape management, were not used frequently in North America before the development of green infrastructure and ecosystems services approaches to planning (Benedict and McMahon, 2006). In their stead, greenways were the most prominent form of landscape-scale investment in green space. Greenways, as defined by Fábos (1995, 2004) and Little (1990), are multi-scaled networks of multifunctional spaces utilising linear and circular landscape features, which are linked to residential and commercial zones to provide access to nature, recreational and educational areas. The use of linear spaces in several locations in the USA reflects an ongoing reimagining of former industrial and transport infrastructure as green space. For example, former railroads and industrial spaces were used to facilitate the Atlanta BeltLine greenway development (Roy, 2015). The redevelopment work of Community Forest partnerships in England could be described as undertaking comparable work, as they continue to attempt to regenerate former industrial sites in urban-fringe locations into multifunctional green spaces (Blackman and Thackray, 2007; Mell, 2010). Framing their value, Julius Fábos, one of the most senior authorities on greenways, listed their functions as follows.

1   Greenways of ecologically significant corridors and natural systems: mostly along rivers, coastal areas and ridgelines; to maintain biodiversity and to provide for wildlife migration and appropriate nature studies.
2   Recreational greenways: where networks of trails and water link land and water-based recreational sites and areas; trails and routes often have scenic quality as they pass through diverse and visually significant landscapes. The recreation focus may be on urban or rural areas and the scale may be local, regional, national or international.

3   Greenways with historical heritage and cultural values: to attract tourists and
     to provide recreational, educational, scenic and economic benefits; to provide
     high-quality housing environments at greenway edges for permanent and
     seasonal housing; to accommodate water resources and flood prevention and
     sensitively located alternative infrastructure for commuting (e.g. bike paths
     within urban areas, recycling of waste and storm water); to offer vehicles of
     expression … among many other possibilities.

(Fábos, 1995, p. 5)

The current developments of the High Line in New York illustrate how
contemporary examples of greenways continue to utilise Fábos's characteristics
as part of a transformative landscape-based rationale for investment in green
space. Moreover, although the delivery of greenway projects could be considered
to add significant economic value to an area, the ethos of these developments is
more nuanced or even multi-layered in terms of meeting socio-economic and ecological needs (Lindsey *et al.*, 2001; Walmsley, 2006). Such variation has enabled
LPAs and developers to work with the inherent spatial flexibility of greenways to
address locally specific socio-ecological issues, as they are less constrained by a
rigid set of geographical, political or development principles compared with other
forms of development (Little, 1990).

One of the main differences between academic – and, indeed, public – thinking
on greenways as opposed to green belts is the way in which they conceptualise
similar landscapes. Green belt discussions often focus on the need to preserve
the countryside from urban expansion (Amati, 2008); they are considered to be
a way of retaining a sense of place and of maintaining the British rural idyll.
However, the conceptual focus of greenway planning, as noted above, looks at re-
establishing a much wider-reaching (and inclusive) value for locations along the
rural–urban interface (Ahern, 1995; Ryan *et al.*, 2006). The use of urban-fringe or
the urban–rural transition zone for some greenway planners is a key component
in achieving spatial diversity and the promotion of equity for people. The linking
of forty-three neighbourhoods associated with the development of the Atlanta
BeltLine greenway project is one example of this (www.beltline.org). However,
it also illustrates a significant difference in how green belt and greenway planning
occur. Although both could be considered to be located in similar geographical
spaces – the urban-fringe/transition zone – green belts are seen, more frequently,
as a way of retaining valuable landscapes by limiting development. Furthermore,
we have seen an exclusionary perspective being associated with several of the
green belt examples discussed in previous chapters. Campaign groups attempting to protect the green belt in the UK might be accused of using an "us vs.
them" rhetoric to situate the value of the green belt. Greenways, however, often
attempt to repurpose former industrial and meanwhile spaces into meaningful and
functional locations for increased use by local communities – for example, in
Indianapolis (*Lindsey et al.*, 2001). Greenways, by contrast, therefore appear to
be more inclusionary for a wider range of users (Little, 1990).

(a)

(b)

Figures 6.2a–b The New York Hudson River greenway (Source: Authors' own)

While there may be merit in working from a protectionist stance with green belt designations if we compare them to the more pluralistic nature of revalorisation of underused spaces, as proposed by greenway planning, the rationale behind them can be challenged. Such a shift in emphasis highlights a significant difference in the basic conceptualisations and utility of these two forms of green-space planning, namely *value*. Valuing landscape resources, as discussed in Chapter 3, is a complicated process and fraught with caveats and subtle inference. To assess effectively what value a landscape holds, be it economic, social or ecological, we must first be able to identify succinctly the constituent parts of a location and then to recognise its wider importance within a socio-economic and political context (Nassauer, 1995). As a consequence green belts, unlike a large proportion of greenways, appear to be more overtly political and to have a direct effect on valuation.

In the UK, those in favour of continuing green belt protection discuss the value of nature, the pristine composition of green belts and the need to retain a semblance of rurality (CPRE and Natural England, 2010). This situates green belt policy in many people's opinion as sacrosanct. It also supports a view that green belts are highly valuable in economic, social and ecological terms – a view we contested in previous chapters. In contrast, greenways have been presented as an attempt to revitalise locations of limited socio-economic and ecological value. The utilisation of former industrial and transport infrastructure is the clearest sign that greenways are, potentially, a far more intuitive form of landscape management, as they are reacting to evolving circumstances rather than maintaining a sixty-year-old doctrine (Little, 1990). There are, however, competing discussions focusing on the positive and negative benefits of greenway development. While they make effective use of derelict or meanwhile spaces, they could also be considered as reinforcing existing socio-economic divisions. In several cities in the USA, such as Indianapolis (Lindsey *et al.*, 2001), investment in greenways has highlighted some of the difficulties encountered by planners in effectively addressing perceived racial and socio-economic differences in access to green and open spaces. In this and other cases, consultation with a range of community stakeholders has enabled these issues to be addressed through effective design and ongoing engagement with local people. Further examples suggest that the decision-making for new greenways may favour locations with lower instances of deprivation or crime (Luymes and Tamminga, 1995; Crewe, 2001). There are also a growing number of voices raising concerns that greenway planning creates an artificial rise in property prices – for instance, around the High Line in New York – which has negative impacts on lower-income residents (Furuseth and Altman, 1991; Berke and Conroy, 2000). The development of the Atlanta BeltLine is one example of a greenway project that is attempting to limit such negative perceptions. It covers 23 miles of derelict former railroad and industrial land, creating an urban ring around downtown Atlanta (Kirkman *et al.*, 2012). One of the desired outcomes of the project is the integration of accessible and multifunctional green and open space that allows free access in all neighbourhoods of

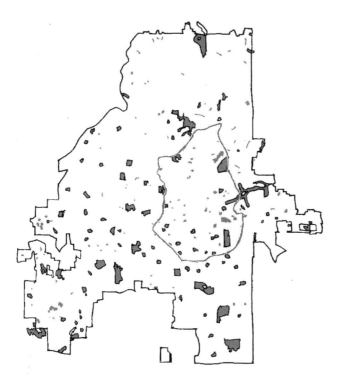

Figure 6.3 Map of the Atlanta BeltLine (Source: Authors' own)

the city (and extending into urban-fringe and county areas in the future). This allows the Beltline to meet local green-space needs in a spatially diverse way, although some commentators have questioned the value of the project if the uplift in property prices does not extend to the communities living close by (Immergluck, 2009; Roy, 2015).

Such diversity in how greenways are valued and conceptualised has provided planners with a range of options for delivering landscape enhancement. This can be contrasted to the more restrictive belt or circular nature of green belt designations used in England and raises interesting questions about the location of green belts compared to greenways and the populations they serve. It also asks whether the benefits of green belts are for the many or for the few; greenways are reported in the academic literature as being for the many (Little, 1990; Hellmund and Smith, 2006).

As discussed previously, there are clear differences between the spatial distribution of greenways and the locations of green belts in the UK and the USA and globally. We can argue that, because of their rural nature, the locations of green belts are potentially more closely aligned with the initial parkway investments in the USA. However, the notion of encouraging people to access

(a)

(b)

Figures 6.4a–b The High Line, New York (Source: Authors' own)

these spaces contrasts to the somewhat exclusionary rhetoric of green belts, although England's National Planning Policy Framework includes a requirement to promote access to these areas (DCLG, 2012). Over time, as parkways gave way to greenways, we have seen a more focused spatial arrangement to greenway planning, more in line with how green belts were designated in the UK. The location of greenways in close proximity to urban and urban-fringe areas illustrates their role in linking urban and rural. Many greenways, therefore, are found along this interface, making use of existing ecological or, in many cases, post-industrial features such as waterways, as in Toronto and Vancouver (Taylor *et al.*, 1995; Fitzsimons *et al.*, 2012). The establishment of a network of spaces crossing this interface is therefore a major design element of greenway planning (McHarg, 1969; Hellmund and Smith, 2006). Working across administrative boundaries has also enabled planners to invest in large-scale landscape projects that bring nature into the lives of people more directly. If we compare this to the development of *The Countryside in and around Towns* agenda promoted in the UK (Countryside Agency and Groundwork, 2005) and the work of England's Community Forest network, we can identify similar attempts to revitalise urban-fringe areas through landscape enhancement (England's Community Forests, 2004; Blackman and Thackray, 2007). Access to greenways for a higher proportion of the population promotes a greater understanding of their use through increased and potentially more long-term exposure. Greenways thus facilitate interactivity and usage as a central characteristic of their development rather than as an added extra.

One of the main differences between the spatial distribution of green belts and greenways appears to be the connecting network aspect of each. While green belts in the UK were designed, predominantly, to encircle urban areas to limit coalescence, greenways are more diverse. Greenway planning could, as a consequence, be considered to be more closely aligned with international green belt provision in places such as São Paulo City Green Belt Biosphere Reserve (Moraes Victor *et al.*, 2004) or the Golden Horseshoe green belt located around Toronto (Taylor *et al.*, 1995; Macdonald and Keil, 2012), which look at the larger spatial aspects of landscape management. Fábos (2004) built on earlier research in the field in his review of greenway practice. He noted, as have other commenters, that the location of greenways was framed by Frederick Law Olmsted and his contemporaries Charles Eliot, H. W. S. Cleveland and George E. Kessler as a mechanism to extend access to green space across urban–rural boundaries. Olmsted's developments in Boston (Massachusetts) and Louisville (Kentucky) also highlight the value of linking urban areas through linear and connective corridors with urban-fringe and rural areas (Little, 1990; Ahern, 1995; Ryan *et al.*, 2002). One of the most significant differences between the spatial distribution of green belts and greenways is the variation in how their use and integration is articulated with wider green networks.

One final aspect that also needs to be considered when discussing green belts and greenways in the UK is the issue of scale. The green belt covers approximately 1.7 million hectares, a not insignificant area of land. This is, of course,

Figure 6.5 Map of the New Delhi green infrastructure network (Source: Authors' own)

split into fourteen or more separate green belts which are correspondingly smaller. Through an analysis of the greenways literature, we can identify areas in the USA such as the New England and Florida greenways, which link ecological networks at a regional and state scale, respectively. Planning at such a scale automatically increases the geographical scope of a greenway, as well as providing opportunities for smaller green infrastructure resources to feed into the wider network (Bueno *et al.*, 1995; Ryan *et al.*, 2006). As a consequence, the size, function and access to greenways can have a significant impact on the socio-economic and ecological value of an area. At a smaller urban/city level, the designing of greenways as cross-boundary resources also allows planners to scale up the strategic objectives and values of investment in green space.

When compared to green belt in the UK, greenway planning in the USA and Canada appears to dwarf, spatially, the size of existing designations. However, comparisons can be made between these investments if we take a proportionality perspective. For example, the green belt around London has played a significant role in limiting coalescence in the South East of England. Its size was therefore representative of the growth projections made for London after 1945 (Hall, 2002; Hall and Tewdwr-Jones, 2010). Comparable investments can be seen in Canada, where the greenway projects in Toronto and Vancouver can be classified as city/ city-region projects providing landscape protection, access to nature and boundaries between urban development and ecological resources. Larger greenways are thus not inherently more effective then smaller sites. What is important is how these investments meet the socio-economic and ecological needs of a specific environment. As a consequence, we might consider that greenways and green belts promote comparable objectives, including protection, interaction and offering an increased value to the urban–rural interface (Hellmund and Smith, 2006; Amati and Taylor, 2010).

Networks of green space

The key difference between greenways and greenspace networks is the scale of the resource base. Greenspace networks are composed of a number of small green spaces which, on account of the breadth of resources, can collectively establish an integrated system with a much greater level of socio-economic and ecological support. Networks of green space therefore provide planners with opportunities to scale up their approaches to landscape management by aligning complementary landscape elements into a collective whole (Matless, 1998). Working from such a perspective enables planners and, to a lesser extent, developers to think more broadly about the role and value of individual spaces and how they can be seen as forming a cohesive and interactive system (Ahern, 2013). Our understandings of greenspace networks can be based on the assumption that resources at the urban– rural interface are not the same, and that they should be managed according to local contexts (Amati and Taylor, 2010).

On account of the diversity of landscape features that can be incorporated into a greenspace network, there is a corresponding increase in the flexibility of what it can look like and where it can be located. Greenspace networks are not restricted to the same rural–urban interface as green belts, although they may, and often do, cover the same geographical area (Ignatieva et al., 2010; Leibenath et al., 2010). This reflects the lack of a specific spatial definition for a "green network" compared with the prescribed understanding of green belts, which may restrict the physical investment of a "belt" or "loop" around urban areas. The flexibility of their composition means that greenspace networks are better able to address ecological issues at a landscape scale. For example, green belts are endorsed as protecting high-quality biodiversity (Zmelik et al., 2011) – a view we discussed in Chapter 3; however, the diverse geographical and ecological nature of a greenspace network is more likely to hold a greater variation and quality of biodiversity because it covers a wider area (Bueno et al., 1995). Furthermore, ecological systems, by their very nature, work more effectively at a landscape scale (Farina, 2006). This view has been embedded into landscape policy – for example, in the European Union's Water Framework

Figure 6.6 Map of the Milan green infrastructure network (Source: Authors' own)

Directive – which explicitly talks about valuing networks on this level (Hering *et al.*, 2010). The European Landscape Convention, with its discussion of network functions for quotidian spaces, goes further. Although some commentators may not consider wider landscape discussions to be appropriate in green belt debates, we might argue that the location of these resources in both "town" and "country" means that landscape-scale debates are actually more appropriate (Amati, 2008; Hansen *et al.*, 2015). Being able to draw on a wider resource base to support ecological and social activities (and to extend the economic value of these resources) offers significant benefits to the planners working in urban-fringe and rural areas.

Greenspace networks also provide planners with a more robust form of spatial continuity than green belts to plan strategically across boundaries. Building on the principles of landscape ecology, networks are an amalgamation of links, hubs and nodes that bring together larger sites with smaller locations linked via linear or circular corridors (Farina, 2006). Where green belts are predominantly circular in nature and encompass urban areas, greenspace networks can cover a much wider area and integrate a broader range of landscape features. They can also be considered to offer a greater level of continuity, as they work with, and are supported by, complementary landscape features – i.e. the regulating and supporting services explored in ecosystem services discussions (Liquete *et al.*, 2015). The practice of linking landscape elements together also increases the likelihood of promoting greater multifunctionality within the environment by offering a range of ecological habitats to promote biodiversity, as well as recreational and economic opportunities. However, as we discussed in Chapter 3, green belts may be viewed in many locations as mono-functional. The use of a network perspective runs counter to this argument, as the integration of a range of resources within a spatially diverse landscape promotes a more dynamic use of the system and an increase in the number of benefits it can deliver (Selman, 2009).

Working at a landscape scale may also have a positive impact in leveraging political and financial support for investment in green-space planning. Although greenspace networks require multi-agency cooperation, there are potentially a greater number of mutual benefits available. The research literature suggests that attempts at coordination between LPAs in green belts can be problematic because of competing protection/conservationist views (Local Government Association and Planning Advisory Service, 2014). However, if a working relationship can be developed between administrations across legislative boundaries, it may be possible to develop a more effective consensus in policy-making, management and funding, as exemplified by the *Cambridgeshire Green Infrastructure Strategy* (Cambridgeshire Horizons, 2011). Unfortunately, to achieve such a position has proved difficult in many places, as competing development objectives often outweigh the collective benefits that can arise through cooperation (Beatley, 2000; Roe and Mell, 2013).

The development of the Atlantic Gateway project in the North West of England is one example of how a multi-scaled approach to development can be aided when LPAs and developers work collectively. Although still in its planning phase, the Atlantic Gateway proposes to utilise the existing landscape resources of the Liverpool–Manchester area, in particular the River Mersey and the area's waterways, as a basis for the enhancement of a landscape-scale network of accessible green and open spaces (www.atlanticgateway.co.uk). This includes the design and implementation of new country parks and woodlands (landscape hubs), greenways and walking/cycling trails (links), and smaller community green spaces (nodes), all of which will form a network of accessible and functional spaces. Greenspace networks also appear to be more accepted in terms of what they can achieve in terms of landscape and urban-fringe management. Because of the spatial diversity that networks offer planners and the environmental sector, they can be used to identify more effective ways to integrate ecological resources into management practices (James *et al.*, 2009). LPAs thus have a wider choice of management, protection and development approaches for landscape networks than is the case with existing green belt designations.

We might also argue that, on a global scale, it is easier to identify a greater number of greenspace networks than green belt designations. "World cities" such as London, Paris and Shanghai have extensive networks of green space that link key residential, commercial, transport and recreational spaces in urban areas to the periphery (Mell, 2016). While these networks provide important social and ecological resources in urban areas, they also serve as key links between and across the urban–rural interface. This allows residents to make the best use of landscape features and affords opportunities for ecological resources to disperse across a wider network of connected habitats, as in Copenhagen (Caspersen and

Olafsson, 2010). In London there is an extensive complex of green spaces which, when assessed alongside the existing green belt, provides a contiguous network of resources, allowing people to move relatively freely between the urban core and the wider countryside. Paris is similar in that the Bois de Vincennes and the Bois de Boulogne both act as green lungs linking the urban core with the peripheral neighbourhoods and rural areas. These two sites act as stepping stones in the greater Paris landscape that (a) limit development to the west and east of the city, (b) provide significant environmental resources, (c) act as centres of formal and informal activity in the urban fringe, and (d) provide green-space hubs linking the city's other green infrastructure and the wider network of green space around Paris (van der Velde and de Wit, 2015).

Finally, there is a need to discuss how people in urban and rural areas view networks of green spaces as compared to green belts. The campaigning literature of some interest groups in the UK appears more likely to declare an interest in green belts, because they are a known entity (CPRE and Natural England, 2010; Sturzaker and Shucksmith, 2011). However, the same level of knowledge, commitment to protection, or use does not seem to be applied to the understanding of greenspace networks. While their specific locales potentially make it easier for local communities to engage with existing green belts, the size, composition and function of networks can be far more difficult to contextualise. Thus, there may be a more limited understanding of how they impact on people's lives and influence the use of the landscape in the rural–urban interface (Blackman and

Figure 6.7 Map of the London green infrastructure network (Source: Authors' own)

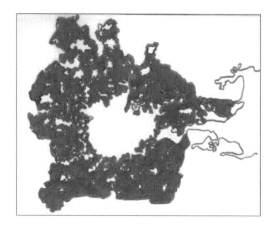

Figure 6.8 Map of the London green belt (Source: Authors' own)

Thackray, 2007). Green belts also have a visible socially constructed value compared to greenspace networks. As a consequence, explaining the composition and value of the latter to stakeholders may actually be more difficult for LPAs. However, if viewed spatially this may be addressed. Greenspace networks could thus be considered to offer more flexibility in terms of administration and provide greater spatial opportunities to integrate the management of different resources as a collective whole because they can draw on a more supportive set of interactive landscape elements.

Green wedges

Located at an intersection between greenspace networks and green belts are green wedges, best exemplified by the "finger plan" of Copenhagen (Denmark). A form of green-space planning that modifies the circular nature of green belts, green wedges promote the integration of green space into urban areas from the periphery inwards (and vice versa). Using linear corridors as an approach to green-space planning, they offer greater continuity of form between what we consider to be green belt, networks of green infrastructure resources and individual sites. The discussion of Copenhagen's use of green wedges highlights the value of adopting a linear approach to investment that transcends neighbourhood boundaries and instead works at a city scale (Beatley, 2000). The benefits associated with the city's green wedges include climate change mitigation, increased recreational space, improved access to nature close to housing, better connectivity for people and biodiversity, and an overarching extension of landscape functionality across the urban–countryside boundary (Caspersen *et al.*, 2006).

Since the first discussion of "green fingers" in its 1948 Regional Plan, Copenhagen has attempted to plan development strategically as a form of "in-fill" between its green wedges. The rationale behind this process has been to make:

1   new development in accessible locations along public transport corridors;
2   the concentration of relatively self-contained new development at specific points along the "fingers";
3   retention of public open space and agricultural land in the "green wedges";
4   allocation of land for allotments and forestry;
5   balancing of population and employment within the metropolitan area.

(Stead, 2000, p. 41)

Points 2 and 3 are potentially the most compatible with the reasons given for supporting green belts in the UK. However, the remaining three principles illustrate how Copenhagen has led the world in the way it has looked to increase the access, connectivity and functionality of the city through the protection of green wedges.

Over time, Copenhagen's 1948 Regional Plan has been revised to extend the spatial reach of the fingers (Stead, 2000). In 1989 it advised that the decentralisation of a number of economic services from the core of the city to peripheral urban centres could be facilitated if they were linked to an expanded finger plan. This proposal made use of a series of additional rings of development (and green space) created around the city to ensure that the spatial continuity of the wedges remained in place (Caspersen and Olafsson, 2010). Moreover, Copenhagen has made efforts to retain its spatial form and has drawn informally on a number of the characteristics discussed by Ebenezer Howard in the garden cities literature. Its use of a radial system of services, infrastructure and green spaces linked to housing and commercial areas could be seen as a "real life" spatial representation of Howard's concepts (Howard, 1898). The longevity of the "finger plan" also reflects the nature of its political support and is, in many ways, comparable to the tenure of green belt designations in the UK. Although the literature suggests that the ideal of green belts appears to be regarded in England as sacrosanct, there have been discussions in Copenhagen examining how best to expand the city and the role that its green wedges should play in this process (Beatley, 2000; Amati and Taylor, 2010). Continuity therefore does not limit debate over the long-term viability of a protected landscape designation, even where policies are known globally. Although the use of green wedges/fingers is deemed successful in international discussions, there remain similarities with green belts in terms of the retention of this form of protection when development objectives threaten to undermine their integrity (Caspersen *et al.*, 2006).

Copenhagen's use of green wedges could be considered as best practice because no other city has been able to replicate the process as successfully. Green belts have a similar appeal in the UK, as despite being challenged as obstructive in planning terms, they have remained government policy for the same period of time (Amati, 2008; Caspersen and Olafsson, 2010). Another key similarity is the level of support that Copenhagen's green wedges receive from politicians and the public. Such a high level of positive support has played a significant role in ensuring they remain in place. Unlike green belts, however, they are also supported by the development sector.

Although Copenhagen may be considered to show best practice in the employment of green wedges, other locations have also utilised this form as an

(a)

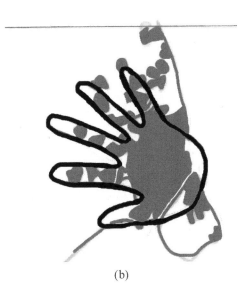

(b)

Figures 6.9a–b Maps of the Copenhagen "finger plan", showing a diagrammatic representation used on the cover of the plan document and the actual built form as it is today (Source: Authors' own)

alternative to green belts. Stockholm (Sweden) has ten green wedges covering 8120 square hectares (21.8 per cent of land cover) that transect the rural–urban divide. These wedges have been promoted in both local and regional planning policy as key locations for the protection of biodiversity and for access to nature and recreational use (Colding *et al.*, 2006). Developed from the remnants of larger areas of forest and green space, the ten wedges are used, in conjunction with a wider green belt area, to manage the expansion of the city. As they cross the boundaries between rural, urban-fringe and urban areas, they offer a variety of opportunities for planners to work with the landscape to ensure that Stockholm retains its multifunctionality. However, it should be noted that the promotion of environmental stewardship, of which conservation and the enhancement of bio-diversity is a priority, has a key influence on the long-term protection offered to landscape resources (Beatley, 2000; Sandström, 2002).

Looking further afield, planners in China have been significant exponents of green wedges in major cities such as Beijing and Shanghai, as well as in smaller cities such as Tianjin (*Li et al.*, 2005). The Shanghai Urban Expo provides one example of where green wedges are key to integrating green spaces in urban areas where developable land is at a premium (Mell, 2016). In their research, Amati and Yokohari (2006) discussed the value in South East Asia of green belt alternatives such as green wedges, stating that they offer a more flexible form of landscape/green-space management that can be more easily adapted to the specific contexts of urban/rural development in rapidly urbanising nations. This, they argued, offers significant benefits, as it allows planners, politicians and developers to use alternative approaches to investment that can lead to a more dynamic form of management and the creation of multiple benefits across urban–rural boundaries (Frey, 2000).

Two examples of this process in China are Jinan (Shandong province, north-west of Beijing) and Nanjing (Jiangsu province, south of Shanghai). In the former, the city's planners developed the "One Ring, Three Green Belts and Nine Wedges Plan" as part of their programme of works to develop Jinan as a national garden city. The role of green wedges was to insert areas of green space into the urban fabric, running from the rural periphery inwards, where a lack of space and development barriers prevented the creation of more formal spaces such as parks. Green wedges were reported by Kong and Nakagoshi (2006) as being of particular importance when planners wanted to connect new residential develop-ments with green areas. The principle promoted by the plan was therefore one of connectivity between people and the landscape to increase socio-economic and ecological values placed on the city's green infrastructure network.

Green wedges have been used in Nanjing to meet comparable needs. Along with a number of alternative forms of green-space planning, such as greenways and green extensions, they have been developed as part of a proposed compre-hensive green-space plan for the city. Nanjing's green wedges have been located to create a 'star shaped urban form' that links the rural hinterland of the city with its urban core through a series of elongated green fingers (Jim and Chen, 2003). Planners have been able to develop a matrix of green spaces that provide suf-ficient areas for use by the urban population that do not disturb other greenfield sites. Moreover, the initial conceptual rationale was a reaction to the development and infill of green and open spaces within the city. Thus green wedges have been used as a protectionist form of planning to ensure that Nanjing retains its green infrastructure resource base.

Green wedges have been proposed as a deliverable and favourable approach to investment in green space. Because of their diversity in size, the locations where they can be developed (or retrofitted) varies in different urban areas. The associated socio-economic benefits they can deliver also make them a viable economic option for many LPAs and developers (Schrijnen, 2000). Furthermore, because, like networks, they offer a way of joining landscape resources across the continuum of urban–rural locations, they help to extend the connectivity over and beyond planning boundaries at a local, city and metropolitan scale (Li *et al.*, 2005).

Strong zoning regulations

In a number of locations there is a very clear distinction between an *urban* and a *rural* area. Such distinctions can become blurred as cities develop and the needs of growth place increasing pressures on existing land-use types. Green belts in the UK fall into this category, as they are urban-fringe or rural locations that are discussed in terms of their need for protection and avoiding the expansion of urban areas (Tewdwr-Jones, 2012). It is difficult to differentiate between these positions unless clear guidelines are in place as to what constitutes an area suitable for development and/or what should be protected. Globally this has been addressed through zoning regulations, most noticeably in North America and East Asia. However, as expansion continues, it has become increasingly challenging to manage some designations – for example, the green belt in Ahmedabad in India – on account of the conflicting influences on LPAs of developers and economic pressures (Mell, 2015b).

In some parts of Asia there are very clear distinctions between urban and rural designations which are enforced via zoning regulations (Yokohari *et al.*, 2000). However, as cities have increased in size, especially mega-cities such as New Delhi and Beijing, there has been increasing pressure to overlook existing landscape protection regulations, including green belts, and allow development. This can, and has, led to a more dynamic and, in some cases, chaotic pattern of land use, as protection for green and open spaces has become more fragmented. This could be seen as a politically motivated process whereby those stakeholders with influence seek to modify zoning regulations to allow development which is beneficial for some but not all members of a community (Robbins, 2012). As a consequence we have witnessed many rural areas being rezoned as "urban". This is becoming increasingly common as new urban extensions emerge, as in Suzhou in China, where the amount of countryside between that city and Shanghai, some 50 miles to the west, is now very limited (Mell, 2016). Furthermore, as more suburbanisation and counter-urbanisation take place, the rapidity of zoning changes as well as the fluidity of this process appears to have increased.

There are a number of high-profile examples of cities using zoning as a mechanism to ensure that development occurs in a structured and strategic manner. In 1939 the Tokyo Green Space Master Plan designated a green belt at a distance of 15 kilometres from the city centre. The aim of this policy was to avoid coalescence between urban areas and to retain individual boundaries through green-space provision (Yokohari *et al.*, 2008). Over time, the rate at which Tokyo has developed has placed increased pressure on its green belt, which has been considered by some commentators as failing to manage coalescence (Watanabe *et al.*, 2008). In South Korea, the city planners in Seoul used green belts as a defensive strategy to limit coalescence between locations. However, there is variation in how people view these resources, with 85 per cent of those surveyed in the city supporting these designations as a mechanism to manage the landscape. However, when they interrogated the data further, Yokohari *et al.* (2000) found that 67 per cent of people who owned land in the green belt disagreed with the zoning regulations. Therefore, even in a city with a (generally) supportive population, the discussion of green belts as a form of land-use management generates a range of alternative views which are influenced by a number of socio-economic variables. It also repeats once more the question we have raised a number of times as to whether restrictive designations can protect the landscape in perpetuity when urban growth is the preferred development strategy.

Also in the Pacific Ring, Melbourne has been an advocate of zoning to support the development of green wedges and green belts for a number of years. First proposed as a mechanism for limiting urban expansion into rural areas, the city's green belts were referred to as 'holding zones' between the rural–urban areas (Buxton and Goodman, 2003). Based on Howard's garden cities principles, Melbourne is distinctive in Australia as it has a long history of green-space policy aimed at protecting the city's environment. Such clear zoning of what is considered urban and rural, with the green belt acting as a buffer, was part of a clear programme of strategic development to moderate expansion. As part of the *Melbourne 2030* strategy, the zoning for green belts was reinforced to ensure urban and 'non-urban' locations remained distinctive (State of Victoria, 2002). However, although the green belt was clearly controlled in the city's development programme between 1996 and 2002, approximately 4000 hectares of land were lost to rezoning for residential development. This illustrates the difficulty of managing designations when political or economic factors start to have greater influence on decision-making. Furthermore, Buxton and Goodman (2003) discussed the impact of a perceived weakness in the legislation concerning zoning changes witnessed during that period. From this, we might argue that, when there is a need to meet local development objectives, a flexible approach to landscape designations can limit the effectiveness of protection at the urban–rural interface.

The experiences of Melbourne with regard to using zoning regulations to protect green spaces, by designating them either as non-development/protection zones or as green belts, has also been seen in North America, even when a flexible approach to management is visible. In the USA, three main types of zoning have been used to manage investment: urban growth boundaries, low-density agricultural zoning, and the purchase of development rights in the rural–urban interface (Daniels, 2010). However, Daniels noted that, although they are viewed as providing a legal structure to enforce management of the landscape, zoning regulations can lack permanence. As a consequence LPAs can modify, and have modified, regulations which previously limited development in green belt areas to meet the needs of residential and commercial development (Freilich, 1999; Daniels, 2010). It has been difficult for LPAs to manage landscape designations in perpetuity because of the fluidity of influences and/or stakeholders who lobby for changes to zoning policy. Daniels does, though, make reference to a series of historical green belt designations that remain influential to zoning arguments in the USA. These include both Olmsted's Emerald Necklace in Boston, which he calls the first green belt in the USA, and Daniel Burnham's plan for Chicago from 1909, which legislated the development of a series of parks and green spaces linking the city with the wider region (Peterson, 2003). This pioneering policy in Chicago led to the creation of the Forest Preserve District in 1914, which gave protection to more than 12,000 hectares of land in and around the city and in the wider Illinois area (Freestone, 2002). The legacy of this can still be seen through the work of the Chicago Wilderness and Forest Preserves agencies, who work with the city's planners (the Chicago Metropolitan Agency for Planning) and LPAs both in the county and in the state of Illinois to maintain landscape protection designations (Mell, 2016). More recently zoning has been used by a number of cities and LPAs in the USA to instigate new green belt designations. Examples are Boulder (Colorado), Baltimore County (Maryland) and Sonoma County (California), each of which has garnered varying degrees of success in both the designation and the management of these resources (Daniels, 2010).

Despite the proposed value of zoning as a mechanism for the protection of the landscape in the USA, such designations are relatively hard to legislate for, as a number of competing influences can attempt to limit their geographical scope or power. Therefore, to ensure that zoning regulations are (a) appropriate (in terms of where and what they are zoning), (b) enforceable and manageable for LPAs, and (c) fair to all stakeholders, there is a need to engage in a constant dialogue between developers, conservationists, planners and people to ensure alignment between influences (Schilling and Logan, 2008; Daniels, 2010). This requires a process of rationalisation to take place that reflects the nature of the landscapes being zoned as well as how best to engage stakeholders, be they landowners, developers or the public, in their management. Landscape professionals and LPA officers in the USA are thus engaged in a process of reviewing the environmental capacity of the green belt resources they are protecting to ensure that they also have a functional social value (Schilling and Logan, 2008).

Summary

How we perceive green belts has a direct influence on how we plan for them. Because of the complexities involved in understanding how their size, location and functionality can address a growing range of rural–urban issues, planners have looked to alternative forms of green-space planning. The alternatives discussed in this chapter illustrate that a number of complementary approaches are available to planners to deliver comparable greenspace and landscape protection to green belt designations. Investment in greenways has a long history in the USA and Canada and has shown how the connectivity of people with the landscape can be used as an overarching theme to link urban and rural green spaces. From their earliest forms in Chicago and Boston (Daniels, 2010), greenways have been reported as offering an approach to landscape protection that crosses administrative boundaries and offers extensive opportunities for people to interact with the landscape. Although they have been used to a lesser extent around the world, they have been seen to facilitate successful human–environmental interactions in and across the urban–rural interface (Amati, 2008). Green wedges and greenspace networks as alternatives to green belts have been held in similarly high regard. For example, greenspace networks provide planners with a range of options to link socio-economically and ecologically important environmental features at a landscape scale. The ability to work across physical, social and administrative boundaries has been seen as a significant benefit offered by such projects (Mell, 2015a). Furthermore, working at a landscape level provides planners with opportunities to integrate a more effective form of systems management for water, ecology/biodiversity and people (Schrijnen, 2000; Jongman and Pungetti, 2004; Leibenath *et al.*, 2010). Green wedges were also discussed in a similar vein, as they are seen as a mechanism for integrating green space into the urban core while retaining links to the periphery or rural areas. Copenhagen was discussed as an example of best practice of this, as the city has both an extensive layout of green wedges and strong political support to maintain them (Caspersen and Olafsson, 2010). These two factors are central to the successful implementation of the principles of connectivity, accessibility and multifunctionality that green space, as well as green belts, is attempting to deliver.

In contrast to the mostly positive green-space alternatives to green belts, zoning offers a more variable interpretation of how protection and management can be undermined. As noted by Daniels (2010), zoning laws change, which leads to a lack of permanence in landscape protection. This has been witnessed around the world as landscape planners in New Delhi, Ahmedabad and Melbourne have reconsidered their zoning regulations in light of development pressures (Buxton and Goodman, 2003; Mell, 2015b). Green belt designations in the UK therefore appear to have greater credence as a result of their longevity, although they are increasingly subject to pressure from housing developers. In China, however, politicians appear to retain significant control over zoning. As the country's cities continue to grow, Chinese politicians and LPA officers are rezoning to accommodate development, yet they are attempting to integrate this within the provision for green and open space (Kong and Nakagoshi, 2006). Shanghai's Urban Expo is one example where such a layered approach to urban development and investment in a variety of green spaces seems to have successfully addressed this challenge – at least in part. Thus, although alternative forms of greenspace planning offer LPAs opportunities, and in many cases solutions, they are also subject to contextual development pressures, such as economic viability, political support, ecological capacity or social needs. Where a strong understanding of environmental and socio-economic systems exists we can see extensive use of greenways, wedges and networks as a way of managing the landscape. Where there is less flexibility in the use of such approaches, zoning can be, and has been, used to manage growth. Unfortunately, because of the lack of permanence of some zoning regulations, it has proven difficult to protect environmental resources in the long term (Schilling and Logan, 2008; Daniels, 2010).

Each of these approaches offers planners a set of alternatives to green belts to manage the landscape. Some have been discussed as being more successful than others. All, however, show that, with a more broadminded understanding of how the urban–rural interface works, different forms of landscape management can be practised. The successful implementation of these approaches may, though, be reliant on the ability of stakeholders to think more innovatively about how best to manage the landscapes in and around urban areas.

References

Ahern, J. (1995) Greenways as a planning strategy, *Landscape and Urban Planning*, 33(1–3): 131–55.

Ahern, J. (2013) Urban landscape sustainability and resilience: the promise and challenges of integrating ecology with urban planning and design, *Landscape Ecology*, 28(6): 1203–12.

Amati, M. (ed.) (2008) *Urban Green Belts in the Twenty-First Century* (Aldershot: Ashgate).

Amati, M., and Taylor, L. (2010) From green belts to green infrastructure, *Planning Practice and Research*, 25(2): 143–55.

Amati, M., and Yokohari, M. (2006) Temporal changes and local variations in the functions of London's green belt, *Landscape and Urban Planning*, 75: 125–42.

Austin, G. (2014) *Green Infrastructure for Landscape Planning: Integrating Human and Natural Systems* (New York: Routledge).

Beatley, T. (2000) *Green Urbanism: Learning from European Cities* (Washington, DC: Island Press).

Benedict, M. A., and McMahon, E. T. (2006) *Green Infrastructure: Linking Landscapes and Communities* (Washington, DC: Island Press).

Berke, P. R., and Conroy, M. M. (2000) Are we planning for sustainable development?, *Journal of the American Planning Association*, 66(1): 21–33.

Blackman, D., and Thackray, R. (2007) *The Green Infrastructure of Sustainable Communities* (North Allerton: England's Community Forest Partnership).

Bueno, J. A., Tsihrintzis, V. A., and Alvarez, L. (1995) South Florida greenways: a conceptual framework for the ecological reconnectivity of the region, *Landscape and Urban Planning*, 33(1–3): 247–66.

Buxton, M., and Goodman, R. (2003) Protecting Melbourne's green belt, *Urban Policy and Research*, 21(2): 205–9.

Cambridgeshire Horizons (2011) *Cambridgeshire Green Infrastructure Strategy* (Cambridge: Cambridgeshire Horizons).

Caspersen, O. H., and Olafsson, A. S. (2010) Recreational mapping and planning for enlargement of the green structure in Greater Copenhagen, *Urban Forestry & Urban Greening*, 9(2): 101–12.

Caspersen, O. H., Konijnendijk, C. C., and Olafsson, A. S. (2006) Green space planning and land use: an assessment of urban regional and green structure planning in Greater Copenhagen, *Geografisk Tidsskrift–Danish Journal of Geography*, 106(2): 7–20.

Colding, J., Lundberg, J., and Folke, C. (2006) Incorporating green-area user groups in urban ecosystem management, *AMBIO: A Journal of the Human Environment*, 35(5): 237–44.

Countryside Agency and Groundwork (2005) *The Countryside in and around Towns: A Vision for Connecting Town and County in the Pursuit of Sustainable Development* (Wetherby: Countryside Agency & Groundwork).

CPRE and Natural England (2010) *Green Belts: A Greener Future* (London and Sheffield: CPRE & Natural England).

Crewe, K. (2001) Linear parks and urban neighbourhoods: a study of the crime impact of the Boston south-west corridor, *Journal of Urban Design*, 6(3): 245–64.

Daniels, T. L. (2010) The use of green belts to control sprawl in the United States, *Planning Practice and Research*, 25(2): 255–71.

DCLG (2012) *National Planning Policy Framework* (London: Department of Communities and Local Government).

England's Community Forests (2004) *Quality of Place, Quality of Life* (Newcastle: England's Community Forests).

Fábos, J. G. (1995) Introduction and overview: the greenway movement, uses and potentials of greenways, *Landscape and Urban Planning*, 33(1–3): 1–13.

Fábos, J. G. (2004) Greenway planning in the United States: its origins and recent case studies, *Landscape and Urban Planning*, 68(2–3): 321–42.

Farina, A. (2006) *Principles and Methods in Landscape Ecology: Towards a Science of the Landscape* (London: Springer).

Fitzsimons, J., Pearson, C. J., Lawson, C., and Hill, M. J. (2012) Evaluation of land-use planning in greenbelts based on intrinsic characteristics and stakeholder values, *Landscape and Urban Planning*, 106(1): 23–34.

Freestone, R. (2002) Greenbelts in city and regional planning, in K. Parsons and D. Schuyler (eds), *From Garden City to Green City: The Legacy of Ebenezer Howard* (Baltimore: Johns Hopkins University Press), pp. 67–98.

Freilich, R. (1999) *From Sprawl to Smart Growth* (Chicago: American Bar Association).

Frey, H. W. (2000) Not green belts but green wedges: the precarious relationship between city and country, *Urban Design International*, 5(1): 13–25.

Furuseth, O. J., and Altman, R. E. (1991) Who's on the greenway: socioeconomic, demographic, and locational characteristics of greenway users, *Environmental Management*, 15(3): 329–36.

Hall, P. (2002) *Cities of Tomorrow: An Intellectual History of Urban Planning and Design in the Twentieth Century* (3rd ed., Oxford: Blackwell).

Hall, P., and Tewdwr-Jones, M. (2010) *Urban and Regional Planning* (Abingdon: Routledge).

Hansen, R., Buizer, M., Rall, E., DeBellis, Y., Davies, C., Elands, B., et al. (2015) *Report of Case Study City Portraits, Appendix: Green Surge Study on Urban Green Infrastructure Planning and Governance in 20 European Case Studies*, http://greensurge.eu/filer/GREEN_SURGE_Report_of_City_Portraits.pdf.

Hellmund, P. C., and Smith, D. (2006) *Designing Greenways: Sustainable Landscapes for Nature and People* (Washington, DC: Island Press).

Hering, D., Borja, A., Carstensen, J., Carvalho, L., Elliott, M., Feld, C. K., et al. (2010) The European Water Framework Directive at the age of 10: a critical review of the achievements with recommendations for the future, *Science of the Total Environment*, 408(19): 4007–19.

Howard, E. (1898) *To-Morrow: A Peaceful Path to Real Reform* (London: Swann Sonnenschein).

Ignatieva, M., Stewart, G. H., and Meurk, C. (2010) Planning and design of ecological networks in urban areas, *Landscape and Ecological Engineering*, 7(1): 17–25.

Immergluck, D. (2009) Large redevelopment initiatives, housing values and gentrification: the case of the Atlanta BeltLine, *Urban Studies*, 46(8): 1723–45.

Jain, A. (2014) NGT to ascertain concretisation of "green belts", *The Hindu*, 30 September, www.thehindu.com/news/cities/Delhi/ngt-to-ascertain-concretisation-of-green-belts/article6460486.ece.

James, P., Tzoulas, K., Adams, M. D., Barber, A., Box, J., Breuste, J., et al. (2009) Towards an integrated understanding of green space in the European built environment, *Urban Forestry & Urban Greening*, 8(2): 65–75.

Jim, C., and Chen, S. (2003) Comprehensive greenspace planning based on landscape ecology principles in compact Nanjing City, China, *Landscape and Urban Planning*, 65(3): 95–116.

Jongman, R., and Pungetti, G. (eds) (2004) *Ecological Networks and Greenways: Concept, Design and Implementation* (Cambridge: Cambridge University Press).

Kirkman, R., Noonan, D. S., and Dunn, S. K. (2012) urban transformation and individual responsibility: the Atlanta BeltLine, *Planning Theory*, 11(4): 418–34.

Kong, F., and Nakagoshi, N. (2006) Spatial-temporal gradient analysis of urban green spaces in Jinan, China, *Landscape and Urban Planning*, 78(3): 147–64.

Leibenath, M., Blum, A., and Stutzriemer, S. (2010) Transboundary cooperation in establishing ecological networks: the case of Germany's external borders, *Landscape and Urban Planning*, 94(2): 84–93.

Li, F., Wang, R., Paulussen, J., and Liu, X. (2005) Comprehensive concept planning of urban greening based on ecological principles: a case study in Beijing, China, *Landscape and Urban Planning*, 72(4): 325–36.

Lindsey, G., Maraj, M., and Kuan, S. (2001) Access, equity, and urban greenways: an exploratory investigation, *Professional Geographer*, 53(3): 332–46.

Liquete, C., Kleeschulte, S., Dige, G., Maes, J., Grizzetti, B., Olah, B., et al. (2015) Mapping green infrastructure based on ecosystem services and ecological networks: a pan-European case study, *Environmental Science & Policy*, 54: 268–280; www.sciencedirect.com/science/article/pii/S1462901115300356.

Little, C. (1990) *Greenways for America* (Baltimore: Johns Hopkins University Press).

Local Government Association and Planning Advisory Service (2014) *Planning on the Doorstep: The Big Issues – Green Belt* (London: LGA & PAS).

Luymes, D. T., and Tamminga, K. (1995) Integrating public safety and use into planning urban greenways, *Landscape and Urban Planning*, 33(1–3): 391–400; www.sciencedirect.com/science/article/pii/016920469402030J.

Macdonald, S., and Keil, R. (2012) The Ontario greenbelt: shifting the scales of the sustainability fix?, *Professional Geographer*, 64(1): 125–45.

McHarg, I. L. (1969) *Design with Nature* (Chichester: John Wiley).

Matless, D. (1998) *Landscape and Englishness* (London: Reaktion Books).

Mell, I. C. (2010) Green infrastructure: concepts, perceptions and its use in spatial planning, unpublished PhD thesis, Newcastle University.

Mell, I. C. (2015a) Green infrastructure planning: policy and objectives, in D. Sinnett, S. Burgess and N. Smith (eds), *Handbook on Green Infrastructure: Planning, Design and Implementation* (Cheltenham: Edward Elgar), pp. 105–23.

Mell, I. C. (2015b) Establishing the rationale for green infrastructure investment in Indian cities: is the mainstreaming of urban greening an expanding or diminishing reality?, *AIMS Environmental Science*, 2(2): 134–53.

Mell, I. C. (2016) *Global Green Infrastructure: Lessons for Successful Policy-Making, Investment and Management* (Abingdon: Routledge).

Moraes Victor, R. A. B., Costa Neto, J. de B., Nacib Ab'Saber, A., Serrano, O., Domingos, M., Pires, B. C. C., et al. (2004) Application of the biosphere reserve concept to urban areas: the case of São Paulo City Green Belt Biosphere Reserve, Brazil, *Annals of the New York Academy of Sciences*, 1023: 237–81.

Nassauer, J. (1995) Culture and changing landscape structure, *Landscape Ecology*, 10(4): 229–37.

Peterson, J. (2003) *The Birth of City Planning in the United States, 1840–1917* (Baltimore: Johns Hopkins University Press).

Robbins, P. (2012) *Political Ecology: A Critical Introduction* (Oxford: Wiley Blackwell).

Roe, M., and Mell, I. C. (2013) Negotiating value and priorities : evaluating the demands of green infrastructure development, *Journal of Environmental Planning and Management*, 56(5): 37–41.

Roy, P. (2015) Collaborative planning – a neoliberal strategy? A study of the Atlanta BeltLine, *Cities*, 43: 59–68.

Ryan, R. L., Fábos, J. G., and Allan, J. J. (2006) Understanding opportunities and challenges for collaborative greenway planning in New England, *Landscape and Urban Planning*, 76(1–4): 172–91.

Ryan, R. L., Fábos, J. G., and Lindhult, M. S. (2002) Continuing a planning tradition: the New England greenway vision plan, *Landscape Journal*, 21(1): 164–72.

Sandström, U. (2002) Green infrastructure planning in urban Sweden, *Planning Practice and Research*, 17(4): 37–41.

Schilling, J., and Logan, J. (2008) Greening the rust belt: a green infrastructure model for right sizing America's shrinking cities, *Journal of the American Planning Association*, 74(4): 451–66.

Schrijnen, P. M. (2000) Infrastructure networks and red–green patterns in city regions, *Landscape and Urban Planning*, 48(3–4): 191–204.

Selman, P. (2009) Planning for landscape multifunctionality, *Sustainability: Science, Practice and Policy*, 5(2): 45–52.

State of Victoria, Department of Infrastructure (2002) *Melbourne 2030: Planning for Sustainable Growth* (Melbourne: State of Victoria, Department of Infrastructure).

Stead, D. (2000) Unstainable settlements, in H. Barton (ed.), *Sustainable Communities: The Potential for Eco-Neighbourhoods* (London: Earthscan), pp. 29–48.

Sturzaker, J., and Shucksmith, M. (2011) Planning for housing in rural England: discursive power and spatial exclusion, *Town Planning Review*, 82(2): 169–94.

Taylor, J., Paine, C., and FitzGibbon, J. (1995) From greenbelt to greenways: four Canadian case studies, *Landscape and Urban Planning*, 33(1–3): 47–64.

Tewdwr-Jones, M. (2012) *Spatial Planning and Governance: Understanding UK Planning* (Basingstoke: Palgrave Macmillan).

van der Velde, R., and de Wit, S. (2015) Representing nature. late twentieth century green infrastructures in Paris, in S. Nijhuis, D. Jauslin and F. van der Hoeven (eds), *Flowscapes – Designing Infrastructure as Landscapes* (Delft: Technical University of Delft), pp. 205–28.

Walmsley, A. (2006) Greenways: multiplying and diversifying in the 21st century, *Landscape and Urban Planning*, 76(1–4): 252–90.

Watanabe, T., Amati, M., Endo, K., and Yokohari, M. (2008) The abandonment of Tokyo's green belt and the search for a new discourse of preservation in Tokyo's suburbs, in M. Amati (ed.), *Urban Green Belts in the Twenty-First Century* (Aldershot: Ashgate), pp. 21–37.

Wu, F. (2015) *Planning for Growth: Urban and Regional Planning in China* (New York: Routledge).

Yokohari, M., Takeuchi, K., Watanabe, T., and Yokota, S. (2000) Beyond greenbelts and zoning: a new planning concept for the environment of Asian mega-cities, *Landscape and Urban Planning*, 47(3–4), 159–71; www.sciencedirect.com/science/article/pii/S0169204699000845.

Yokohari, M., Takeuchi, K., Watanabe, T., and Yokota, S. (2008) Beyond greenbelts and zoning: a new planning concept for the environment of Asian mega-cities, in J. Marlzuff, E. Shulenberger, W. Endlicher, M. Alberti, G. Gradley, C. Ryan, et al. (eds), *Urban Ecology: An International Perspective on the Interaction between Humans and Nature* (New York: Springer International), pp. 783–96.

Zhang, T. (2005) Uneven development amongst Shanghai's three urban districts, in L. Ma and F. Wu (eds), *Restructuring the Chinese City: Changing Society, Economy and Space* (Abingdon: Routledge), pp. 124–39.

Zmelik, K., Schindler, S., and Wrbka, T. (2011) The European green belt: international collaboration in biodiversity research and nature conservation along the former iron curtain, *Innovation: The European Journal of Social Science Research*, 24(3): 273–94.

7   Conclusion

Introduction

This book is titled *Green Belts: Past; present; future?* Chapter 2, with a review of the evolution of policy and practice of green belts in the UK, concentrated on the past. Chapters 3, 4 and 5 are concerned with the present, dealing respectively with an analysis of the impacts of green belts, an examination of their characteristics, and a discussion of some examples of comparable practice from around the world. Chapter 6 began a look towards the future by identifying some alternatives to green belts. Here, following a brief summary of the preceding chapters, we conclude this discussion of the future and focus particularly on that provocative "?" in our title.

As we will go on to elucidate, we believe the question of the future of green belts should be very much an open one. As we noted in our opening chapter, there is a tendency for public discussions on green belts to default to an argument between two fairly extreme positions: on one side, more or less unswerving support for green belts, often on the basis that they are essential to preserve the countryside (such as CPRE, 2016a); on the other, a view that green belts are symptomatic of a planning system that stifles economic development, usually inspired by a neoliberal ideology (for example, Papworth, 2015). It is important here to stress, as we have done several times, that we firmly believe in public-sector-driven planning. We do not share the neoliberal view that planning itself is problematic, or that the best way to achieve progress is to remove the "shackles" that bind entrepreneurs. Alternatively, we want to argue that a pro-planning position is compatible with a robust critique of green belts and that a different approach to land (and landscape) management may be more effective in delivering benefits for society. In this chapter we will expand upon and synthesise these arguments. Before that, however, we present summaries of our discussions to this point.

Green belts: the past

Chapter 2 of this book comprised a historical review of the development of ideas, rhetoric and policy around green belts in the UK. We discussed the earliest evidence of these, found in ancient India and the Roman Empire, and the reference to something very like green belts in the Bible. In relation to (what is now) the

UK, concern over the growth of cities (specifically London) in the fifteenth and sixteenth centuries led to various arguments for the limitation of urban development (Evelyn, [1661] 2011); Anonymous, 1956; More, [1556] 1999). But it was the massive rate of urbanisation which resulted from the Industrial Revolution in the eighteenth and nineteenth centuries, the resulting high population density in the towns and cities, and the consequent spread of disease which led to a broader level of interest. Various royal commissions and other bodies investigated these issues, and they and others sought viable solutions.

One type of solution which swiftly came to have a profound influence was to identify new models of urban development. James Silk Buckingham is credited with one model, which was part of the inspiration for the most famous example, Ebenezer Howard's garden cities. This mild-mannered clerk published his keystone text, originally entitled *To-Morrow: A Peaceful Path to Real Reform*, in 1898 and reissued it four years later as *Garden Cities of To-Morrow*. These books are widely acknowledged as being the most influential texts in relation to what emerged as the discipline and profession of town planning. Howard's garden city was to be 'as much "a city *in* a garden" as "a city *of* gardens"' (Osborn, 1946, p. 167), with the "gardens" forming a belt around the city. As noted by others, notably Peter Hall (Hall *et al.*, 1973; Hall *et al.*, 2003), Howard did not envisage single isolated garden cities but, rather, a linked cluster of them, necessary as a result of population growth, in what he called in the original 1898 volume the 'Social City'. This idea did not survive even until the 1902 reissue of the book and was unequivocally not part of what emerged as green belt policy.

The idea of using a belt or garter of green to limit the growth of cities, initially and perhaps most importantly London, gradually took hold during the first half of the twentieth century, both in the UK and internationally. The first attempt to legislate for this was the 1938 Green Belt Act, which gave statutory weight to the process that had been ongoing for some years, whereby London County Council bought land in and around the city to preserve it from development. Very shortly thereafter came the Second World War, and with it a recognition that the UK was vulnerable to attack from Europe because the population was concentrated in a small number of large urban areas (London of course being most significant); there had also been a substantial loss of agricultural land to development, increasing the reliance of the country on imported food. As a consequence, four reports were undertaken during and immediately after the war on various planning-related matters (the Barlow, Scott, Uthwatt and Reith reports), which made various recommendations on issues such as green belts, the relocation of the industrial population, and other questions of national importance. These reports had a major influence on the 1947 Town and Country Planning Act and related legislation, which set the parameters of the English planning system that remain broadly in place today. Also influential at this time were the various advisory plans produced by Patrick Abercrombie – then a professor at University College London, a member of the Barlow Commission and a founder member of the Council for the Preservation of Rural England (now the Campaign to Protect Rural England). The plans he produced, most famously for London, generally promoted green belts as a prominent feature.

All the preceding fed into the first nationwide attempt to introduce green belts through the 1955 Ministry of Housing and Local Government Circular 42/55, which was sent to all local planning authorities. The circular (reproduced in full in Box 2.1) identifies 'the importance of checking the unrestricted sprawl of the built-up areas ... [through] the formal designation of clearly defined Green Belts' (MHLG, 1955, p. 1), sets out constraints on development in such green belts, which remain largely unchanged today, and specifies aims and objectives for green belts, which have since been added to incrementally.

The story of green belts since this advice is one which features, perhaps uniquely in the realm of British politics, a continuing and largely undiminished rhetorical consensus from national (and most local) politicians that they are committed to maintaining green belts in the form in which they were designated from 1955 (the process of formal designation continuing until, in some cases, the 1980s). It is hard to think of another national policy which has not changed or been "reformed" in some quite substantial way since 1955, but the principles contained in Circular 42/55 remain largely in place today, sixty years later, despite attempts by at least two of the ministers in charge of green belts to change them more substantially – attempts that were rebuffed by, among others, the CPRE. Perversely, however, those sixty years have also seen constant minor change both to what have variously been called the objectives, purposes, roles and aims of green belts (see Table 2.2) and to the physical area of green belts, with deletions and expansions taking place in different places at different times (Elson, 1986; Ward, 2004). This is an example of what some see as the tendency on the part of the state in the UK to seek to minimise change and stress continuity in policy (Foley, 1963); it continues today, with the current government emphasising support for the green belt, while the CPRE bemoans plans for 275,000 homes thereon (CPRE, 2016a).

The end result of these contradictions is that policy-makers, at both national and local level, conscious of the need for new development but also terrified of the political consequences of irking the CPRE and their allies, find themselves jumping through various rhetorical hoops to justify "exceptions" to policy. At the same time, the public continues to believe that green belts are largely unchanged, valuable and continually protected. This is in no small part due to misunderstandings of what they have actually achieved and what their characteristics are, which we discussed in Chapters 3 and 4.

Green belts: the present

The impacts of green belts

Chapter 3 began by setting out what we could and could not assess. For various reasons it is impossible to be certain about the impacts of green belts, so we brought together evidence on their *apparent* impacts, both in the UK and elsewhere, and attempted to synthesise and evaluate this evidence. Our conclusions were that green belts appear to have had quite a substantial effect on some aspects of urban development but that the evidence is much less clear on others.

There is broad agreement, for example, that they have succeeded in limiting physical urban sprawl, thus achieving what Circular 42/55 specified as their initial aim. What green belts have been much less effective at doing, however, is achieving *functional* containment of urban areas – so-called leapfrogging has occurred in some areas, with development continuing beyond the green belt, at the same time as car ownership has risen, leading to large increases in the amount of commuting and the distances travelled.

This commuting clearly has an environmental impact, but the overall impact of green belts on the traditional three components of "sustainable development" (environmental, social and economic) is far less clear. Perhaps without green belts these trends of commuting would have been worse, and other environmental benefits, such as the conservation of valuable habitats, need to be weighed in the balance. The economic and social effects are also mixed, with some arguing that green belts have negatively impacted on the economy through businesses relocating to other countries because they cannot expand in the UK (Evans & Hartwich, 2007) – this is, however, a minority view. What is a more broadly accepted position is that green belts, and other planning constraints, have increased house prices, both in the UK and elsewhere. Many studies have looked at this question, with most concluding that there is an impact for the fairly basic economic reason that scarcity of a "commodity" (in this case land, and subsequently houses built upon that land) drives up the price of that commodity. The scale of this influence, and the consequences of partial or complete removal of the green belt, is much more contested; estimates of the rise in house prices range from 52 per cent (Hilber and Vermeulen, 2010) to 1.2 per cent (Bramley, 1993b, 1993a), with the other effects of any removal again needing to be weighed in the balance.

The final consequence to mention here is that while, in utilitarian terms, it is not easy (or perhaps impossible) to say whether the aggregate costs of green belts outweigh their benefits, it is perhaps easier to say that these costs and benefits have been unequally distributed, with the costs falling on the less well off and the benefits on the wealthier. So we would argue, as others have done (cf. Hall *et al.*, 1973), that green belts have been regressive in nature.

The characteristics of green belts

Chapter 4 focused on three things. Firstly, it explored the actual size, the location of each designation and the variation in physical landscape that characterises green belts. Our key finding here was that the characteristics of green belts vary hugely from place to place. In some locations they are rich in biodiversity, in others they provide an essential local resource for recreation, and in yet others they protect landscapes that are commonly agreed to be worth protecting. In many, however, they do none of these things (CPRE and Natural England, 2010). In the second part of Chapter 4 we looked at the questions of *value* and *quality* attached to green belts. We considered differences between the *quality* of green belt land, which can be measured to some extent objectively, and the *value* attached to it by those who live near to (or far from) it, which is a much more subjective evaluation. We linked the latter to the so-called pastoral tradition in the UK and the tendency, observed by others, to romanticise the countryside as offering a "rural idyll" (Matless, 1998; Woods, 2005). We asked whether it is possible to continue to rely on historical understandings of green belts to shape how development is controlled today. Moreover, we discussed the complexities of aligning "size" and "value" in terms of identifying the values of green belts in the UK. Finally, in this chapter we brought all these elements together and suggested that the green belt is as much imaginary as it is real. How we consider cultural interpretations of green belts may therefore give us a more nuanced understanding of the factors which influence valuation in policy, practice and perceptions. This both suggests a need to reconsider whether green belt remains an aspect of planning policy that is fit for purpose and illustrates that such a reconsideration is fraught with difficulty – humans are not necessarily purely rational decision-makers, and the powerful appeal of a green belt that ties into our long history of romantic pastoralism is not something that can be easily challenged. If, however, such pastoralism is to be overcome (or at least challenged), what lessons from other parts of the world might be applicable to the UK?

Green belts and similar policies around the world

In Chapter 5 we examined green belts and similar policies in Europe, North America, Asia and Australia. We noted that, despite the differences in context between many (if not all) of our case studies and the situation in the UK, there were lessons to be learned. Some of the cities cited had based their own green belt on the example of the UK (such as Melbourne and Tokyo), but it was possible to identify commonalities in experience whether or not this was the case. For example, in China, which clearly has a different approach to planning (and many other things) from the UK, the move from a state-led and top-down development model to a decentralised and private sector-led approach had resulted in a 'false belief' (Yang and Jinxing, 2007, p. 293) that the state could control urban growth. We would argue that Abercrombie's advisory plans and the assumptions inherent in the 1947 Town and Country Planning Act are predicated on a similar belief, which was rendered false in the following years by changes to planning and development processes and the scale of urban development. This suggests the need to reconsider earlier assumptions when assessing the appropriateness of policies such as green belts.

In other cases – for example, Melbourne and the Netherlands – the approach to large-scale green-space management (through green wedges and the Green Heart respectively) has been in place for many years, so has become part of long-term city management and, in the case of the Netherlands, part of planning doctrine. This has benefits in that it is widely understood by planners, politicians and the public alike, but it is problematic in that it becomes difficult to reform. As with the UK green belt, the Dutch Green Heart is perhaps immune from rational reform.

If reform *is* possible, then we can find pointers for the direction in which we might move from some international examples. The first is that scale matters – smaller sites or areas of green belt are easier to implement and manage, whereas larger green belts can garner greater support from the public and others and allow the management of development at a (city-) regional scale, rather than risking overspill and other interlocal effects. The scale at which green belts are going to operate should dictate how they are used. Secondly, the most successful examples of green belt or similar policy are those where positive uses are emphasised, often in terms of recreation and leisure, which are themselves linked to broader socio-economic and ecological benefits and the promotion of green and open spaces as multifunctional. This helps build and consolidate public support for their continued existence. Thirdly, there is a need for flexibility, as comparisons from the USA show: in California, where urban growth boundaries have been more rigidly enforced, house prices have risen as a result, whereas, in Portland, authorities have used other tools to maintain land supply and keep house prices comparatively affordable.

Chapters 3, 4 and 5 together tell a story about difference and nuance in how green belts work, what impacts they have on house prices and urban density, their characteristics, and the influence of people within the landscape. This is true at a range of scales – nationally (green belts in some countries being more effective than others on account of the specificities of their context and management), regionally (London and the South East having such a profoundly different housing market than the rest of England and the UK) and locally (all green belts having areas of variable quality). Yet all of the green belt across England enjoys the same level of protection from development, a level that has remained in place since 1955. Does this make sense in planning, landscape and social terms? We would argue that it does not, and that reform is needed. But what might this reform entail?

Green belts: the future (?)

Alternatives to green belts

Chapter 6 considered a range of alternative approaches to green belts which might be applicable in the UK, based on three questions.

1   Is it possible to replace green belts with anything that has similar characteristics?
2   What evidence is there of tools available that might achieve similar (positive) impacts to green belts?
3   Are there practical combinations of mechanisms that might work in the UK?

The alternatives considered included zoning of areas as protected green space; greenways; greenspace networks; and green wedges. There are advantages and disadvantages to each. Zoning, in common use around the world but not, at this time, widely used in the UK, has the benefit of being simple, allowing all users to understand clearly what is proposed. It is, however, very easy to amend, which, while building in the kind of flexibility we have identified as being important, also means that important green spaces can be easily lost to development. In many of the arguments against changing green belt designations, added flexibility in terms of their protection is seen as the first step to a lack of permanence. We identified greenways – linked multifunctional green spaces which use landscape features – as being more progressive in many ways than green belts. Their emphasis is often on equality and community use (as much as it is on limiting urban growth), and they are similarly often deliberately located nearer to less wealthy areas, whereas green belts tend to be more accessible to wealthier households. Urban greenways also tend to be smaller in scale, thus not permitting the landscape-level management which is possible with green belts, although some, including the Boston Emerald Necklace (cf. Fábos, 2004) and more historical parkways, cross administrative and physical boundaries. Greenspace networks, by contrast, operate at the landscape scale and are multi-sectoral in nature, the combination of which allows better cross-boundary planning. They are also more flexible than green belts, but this again can be a weakness; as they may be perceived as being less focused and more diffuse, it can be more difficult to build public support for them. However, strong legislative backing for landscape designations can, and does, ensure protection even where competing development objectives exist.

Finally, green wedges sit at the intersection between greenspace networks and green belts. In some areas, including Melbourne and Copenhagen (the latter via the "finger plan"), supported by the public and, in the latter case, the development sector, they have become part of planning doctrine. They allow the linking of green-space resources in a linear way and provide recreational access in a reasonably equitable way throughout urban areas. As we discussed in Chapter 2, following the introduction of Circular 42/55, the TCPA wrote to the Minister for Housing and Local Government with a series of questions and published in its journal that letter and the minister's response. The response indicated that the minister agreed with the TCPA that 'For large conurbations the green belts will be at some distance from the centre. Therefore it is important to preserve green wedges' (Sandys, 1956, p. 152). Sadly, this agreement did not translate into meaningful action, so the opportunity to preserve wedges of green space, easily accessible to more of the population, was lost. But it is not too late to remedy the situation – new green spaces, and the linking of existing resources, are being introduced to urban areas all over the world, and there is scope to do more of the same in the UK's cities.

The need for a balanced view

As we noted in Chapter 1 and the introduction to this chapter – and as we hope has been apparent to you as the reader throughout this book – green belts arouse strong opinions. Individuals and groups tend to be strongly in favour of preserving them or strongly in favour of removing them, in full or in part. The latter position tends to flow from an ideological stance that we would categorise as neoliberal and is sometimes set out in papers emanating from free-market think-tanks

(Papworth, 2015). While such papers draw on evidence that the authors no doubt believe is entirely robust, the influence of the neoliberal position arguably plays a strong role in their findings: it would be something of a surprise if a free-market organisation was to advocate stronger rather than weaker land-use planning control. Conversely, those lobbying for the preservation of green belts would deny they are influenced by any such ideology. Indeed, the CPRE, perhaps the pre-eminent body arguing for the retention of green belts, emphasises its political neutrality, stressing that it 'has no vested interests ... [we] are politically independent. We make decisions with the head not the heart ... our wide remit means we have to consider the whole of the country – rural and urban – when creating policy' (CPRE, 2016b). Our position, however, is that, while the CPRE may have no political affiliation, the arguments made by this and other organisations in favour of blanket retention of green belts are driven by a predetermined position as much as the anti-green belt arguments of the free-market ideologues. That position, as we pointed out in Chapter 4, draws on a deep well of romanticism where the English countryside is concerned – something that is intrinsically important and in need of protection for its own sake. This is as immune from rational argument as is the "free market good, state control bad" position adopted by others.

We have tried in this book to avoid taking predetermined positions or being unduly influenced by ideology and to present the evidence we have gathered in a balanced way. It is for the reader to determine whether or not we have succeeded. In these final pages, drawing on all the evidence we have discussed so far, we want to present our case for reform of the green belt in a way that will protect high-quality landscapes, enhance biodiversity, enable more access to green spaces for recreation and, most importantly from our perspective, do so in a way that reduces (or at the very least does not increase) inequalities between groups in our society. Each of us will have our own view on the interactivity of rational planning arguments for and against green belts, with evidence of their value and functionality and the emotional responses used to frame debates. It is further necessary to consider political and temporal changes in the perceived need for development and the complexity of balancing the realities of planning for and managing green belts.

To set the context for this we wish to draw upon two quotations we came across during our work on this book. The first, written over sixty years ago, is to us as good a description as one could hope for of what planning should be about (everywhere in the world); the second is a concise description of the British planning system that remains as true today as it was twenty years ago, and is the context within which our final words sit.

> Town and country planning is not an end in itself; it is the instrument by which to secure that the best use is made of the available land in the interests of the community as a whole. By nature it cannot be static. It must advance with the condition of society it is designed to serve.
>
> (Uthwatt, 1942, p. 12)

British land use planning is a system of response to change. It operates by guiding a predominantly private development market, and it is a political system.

(Elson, 1986, p. xxiv)

What might an alternative to green belts in the UK look like?

As the quotation above by Elson suggests, reform to green belts has to take place within, on the one hand, a political context, which makes this a difficult and contentious topic, and, on the other, a development context that relies mostly on the private sector. Therefore, while we might like to see, for example, a return to some of Ebenezer Howard's original ideas in the form of a substantial programme of new garden city construction across England, this is extremely unlikely to happen in the short or, indeed, the medium term; in the absence of regional planning or a will on the part of government to reinvigorate development corporations, neither the planning nor the delivery of garden cities (or new towns) on a large scale is feasible. We therefore focus on what we consider to be more realistic proposals in the current context.

The first is to embrace the argument put forward by "Tom Pain" in the journal of the Town and Country Planning Association. This pseudonymous writer shares our view that green belts were not originally conceived as a tool to limit urban containment – rather, their function was to provide access to green space 'as an integrated resource for the people of both the country and the city' (Pain, 2015, p. 157). Pain proposes a threefold solution to facilitate this access: firstly, opening up all green belt in the same way as most land in Scotland, where the population has the "right to roam"; secondly, creating a new body to manage all green belt land and 'promote the enhancement of biodiversity, community access, community farming, and forestry activities' (ibid.); and, thirdly, promoting, through planning policy, 'small-scale food production and low-impact living and low-impact renewables' (ibid.). These are no doubt radical solutions, some of which are more politically palatable than others. The second, in particular, appears to us to be a policy with which it is hard to disagree and one that could be rapidly implemented, especially if an independent yet inclusive body could be created to oversee any future changes.

The second of our proposals is to suggest that the enhancement of green belt access is accompanied by a comprehensive review of all green belt land in England as part of the process of local plan preparation. We would advocate the use of the six objectives of green belt specified in the 1995 edition of *Planning Policy Guidance 2* (PPG2):

- to provide opportunities for access to the open countryside for the urban population;
- to provide opportunities for outdoor sport and outdoor recreation near urban areas;

- to retain attractive landscapes, and enhance landscapes, near to where people live;
- to improve damaged and derelict land around towns;
- to secure nature conservation interest; and
- to retain land in agricultural, forestry and related uses.

(DoE, 1995, para. 1.6)

We suggest that, where areas of green belt are not, and have no realistic prospect of, meeting any of these objectives, they are considered for deletion. We do not consider an overall reduction in the area of green belt to be necessarily a bad thing, but this will be locally contentious. To reduce opposition to this process, therefore, local authorities could be required/encouraged to add to the green belt the same amount of land they wish to delete (either in their own area or in an adjoining local authority through the 'Duty to Cooperate' introduced by the 2011 Localism Act).

Our third proposal is that, to increase access to green space, the provision of green wedges should become a major focus of planning policy. As noted above, it was intended by the originator of Circular 42/55 that green wedges would be provided alongside green belts to ensure that those living nearer the centre of an urban area could access green space, but this was never followed through. The international evidence suggests that green wedges can provide a valuable resource to meet all of the objectives specified for green belts in PPG2, and can do so in a far more equitable way than green belts. It also emphasises that, with more proactive and forward-thinking planning, green spaces can be successfully integrated into urban areas and be protected in the long term.

Finally, we suggest that the time might be right to enter into a renewed national debate focusing on the local and larger-scale values of green belts in the UK. Through such discussions it may be possible for central and/or local government to rethink the sacrosanct nature of green belts, and the ideologies which support their protection, to assess whether they remain fit for purpose. This may provoke strong opposition, but there is an unanswered question as to why, despite the dynamic nature of planning and development in the UK, only this one post-Second World War planning policy remains largely unchanged.

Current emphasis in planning policy is on the facilitation of "sustainable development", which is defined as 'positive growth – making economic, environmental and social progress for this and future generations' (DCLG, 2012, p. i). However, as we discussed in Chapter 2, there is some evidence that opposition to development in the green belt affected both local and national elections in England in 2015. This continued in 2016 both in the London mayoral election, where the (ultimately successful) Labour candidate, Sadiq Khan, was accused of planning to 'Khan-crete them [green spaces] over' (Hill, 2016), and in the local elections, where single-issue anti-development groups stood for election and won seats in several areas of the South East (Gardiner, 2016) and UKIP influenced more mainstream politicians to move away from green belt development (Sell, 2016).

The problems inherent in relying on the profit-driven private sector to meet the pressing need for additional housing (and other forms of development) within urban areas heavily constrained by green belts and similar mechanisms are not going away. A more intelligent approach that acknowledges the variable quality of green belt and supports appropriate development on lower-quality areas, while replacing this with higher-quality green space that is closer to where most people live, seems to us to be a logical and sensible way forward.

Summary

Throughout this book we have asked a series of basic, yet important questions. We have asked what green belts are, what they do, whether they still deliver the functions they were set up to achieve, and if they remain fit for purpose. The preceding chapters have explored these issues, and we have perhaps raised more questions than answers. We are calling not for revocation, or large-scale deletions, of the green belt but for a more inclusive and evidence-based discussion of the current ecological functionality and socio-economic value of green belts. We would argue for a considered rethinking of their value in an era where protectionist policies are increasingly under pressure from those advocating for development. We also argue that the maintenance of green belt designations is a valid form of landscape planning where socio-economic and ecological benefits outweigh their costs. A point we have returned to throughout is that we feel there should be a reconsideration of the almost universally accepted value placed on green belts.

The sixty-plus-year history of green belts as an official planning tool of governments in the UK has shown an incredible level of resilience in the face of population growth, increased mobility and wealth, and continued calls from some to relax the policy. This is an impressive achievement for any approach to planning. However, we believe that it may be time to reconsider the sacrosanct nature of green belt policy and bring discussions into line with contemporary notions of planning as a dynamic, fluid, responsive and participatory process. If, as David Lowenthal (1985) believed, drawing upon L. P. Hartley before him, that the past is a foreign country where planners, developers and the public did things differently, then green belts appear to be the single policy exception that proves this rule. We know of the successes and failures of the past, we understand the development needs of today, and the future provides us with opportunities to think, like Ebenezer Howard, more profoundly about how to move forward. We have an opportunity to reflect upon the value placed on green belts, their utility for people, and the ways in which the policy as a whole, and individual areas of green belt specifically, might adapt to meet future development needs. If we can achieve such a shift in approach, then we may move beyond the ideologically distant "for" and "against" positions that underpin current green belt discussions to a more nuanced and meaningful form of debate and, hopefully, green-space management.

References

Anonymous (1956) Ye olde English green belt, *Journal of the Town Planning Institute*, 42: 68–9.

Bramley, G. (1993a) The impact of land use planning and tax subsidies on the supply and price of housing in Britain, *Urban Studies*, 30(1): 5–30.

Bramley, G. (1993b) Land-use planning and the housing market in Britain: the impact on housebuilding and house prices, *Environment and Planning A*, 25(7): 1021–51.

CPRE (2016a) Green belt undersSiege: 2016, briefing, www.cpre.org.uk/resources/housing-and-planning/green-belts/item/4276-green-belt-under-siege-2016.

CPRE (2016b) Our vision for the countryside, www.cpre.org.uk/about-us/our-vision.

CPRE and Natural England (2010) *Green Belts: A Greener Future* (London & Sheffield: CPRE & Natural England).

DCLG (2012) *National Planning Policy Framework* (London: Department for Communities and Local Government).

DoE (1995) *Planning Policy Guidance 2: Green Belts* (London: HMSO).

Elson, M. J. (1986) *Green Belts: Conflict Mediation in the Urban Fringe* (London: Heinemann).

Evans, A., and Hartwich, O. (2007) *The Best Laid Plans: How Planning Prevents Economic Growth* (London: Policy Exchange).

Evelyn, J. ([1661] 2011) *Fumifugium MMXI: A 21st Century Translation of a 17th Century Essay on Air Pollution in London, Sent to King Charles II by the Writer John Evelyn* (London: Environmental Protection UK).

Fábos, J. G. (2004) Greenway planning in the United States: its origins and recent case studies, *Landscape and Urban Planning*, 68(2–3): 321–42.

Foley, D. L. (1963) *Controlling London's Growth: Planning the Great Wen 1940–1960* (Berkeley: University of California Press).

Gardiner, J. (2016) Why local elections could be influenced by housing hostility, *Planning*, 22 April 2016, p. 8.

Hall, P., Gracey, H., Drewitt, R. & Thomas, R. (1973) *The Containment of Urban England*, Vol. 2 (London: Allen & Unwin).

Hall, P., Hardy, D. & Ward, C. (2003) *To-Morrow: A Peaceful Path to Real Reform. Original Edition with Commentary* (London: Routledge).

Hilber, C., and Vermeulen, W. (2010) *The Impact of Restricting Housing Supply on House Prices and Affordability* (London: HMSO).

Hill, D. (2016) London mayor race: Zac Goldsmith is being slippery about the green belt, *The Guardian*, 23 April, www.theguardian.com/uk-news/davehillblog/2016/apr/23/london-mayor-race-zac-goldsmith-is-being-slippery-about-the-green-belt.

Howard, E. (1898) *To-Morrow: A Peaceful Path to Real Reform* (London: Swann Sonnenschein).

Howard, E. ([1902] 1951) *Garden Cities of To-Morrow*, ed. F. J. Osborn, intro. L. Mumford (London: Faber & Faber).

Lowenthal, D. (1985) *The Past is a Foreign Country* (Cambridge: Cambridge University Press).

Matless, D. (1998) *Landscape and Englishness* (London: Reaktion Books).

MHLG (1955) *Green Belts*, Circular 42/55 (London: HMSO).

More, T. ([1556] 1999) *Utopia* (Boston: Bedford/St Martin's).

Osborn, F. J. (1946) *Green-Belt Cities – the British Contribution* (London: Faber & Faber).

Pain, T. (2015) Ruskin was right about the green belt, *Town and Country Planning*, 82(4): 156–7.

Papworth, T. (2015) *The Green Noose: An Analysis of Green Belts and Proposals for Reform* (London: ASI (Research) Ltd).

Sandys, D. (1956) Green *Belts: Minister's Reply, Town and Country Planning*, 24: 151–3.

Sell, S. (2016) How UKIP has affected planning in areas where it has a strong presence, *Planning*, 22 April 2016, p. 9.

Uthwatt, A. A. (1942) *Final report [of the] Expert Committee on Compensation and Betterment* (London: HMSO).

Ward, S. K. (2004) *Planning and Urban Change* (2nd ed., London: Sage).

Woods, M. (2005) *Contesting Rurality: Politics in the British Countryside* (Aldershot: Ashgate).

Yang, J., and Jinxing, Z. (2007) The failure and success of greenbelt program in Beijing, *Urban Forestry and Urban Greening*, 6(1): 287–96.

Index

Taylor & Francis eBooks

Helping you to choose the right eBooks for your Library

Add Routledge titles to your library's digital collection today. Taylor and Francis ebooks contains over 50,000 titles in the Humanities, Social Sciences, Behavioural Sciences, Built Environment and Law.

Choose from a range of subject packages or create your own!

Benefits for you

>> Free MARC records
>> COUNTER-compliant usage statistics
>> Flexible purchase and pricing options
>> All titles DRM-free.

Benefits for your user

>> Off-site, anytime access via Athens or referring URL
>> Print or copy pages or chapters
>> Full content search
>> Bookmark, highlight and annotate text
>> Access to thousands of pages of quality research at the click of a button.

REQUEST YOUR **FREE** INSTITUTIONAL TRIAL TODAY

Free Trials Available
We offer free trials to qualifying academic, corporate and government customers.

eCollections – Choose from over 30 subject eCollections, including:

Archaeology	Language Learning
Architecture	Law
Asian Studies	Literature
Business & Management	Media & Communication
Classical Studies	Middle East Studies
Construction	Music
Creative & Media Arts	Philosophy
Criminology & Criminal Justice	Planning
Economics	Politics
Education	Psychology & Mental Health
Energy	Religion
Engineering	Security
English Language & Linguistics	Social Work
Environment & Sustainability	Sociology
Geography	Sport
Health Studies	Theatre & Performance
History	Tourism, Hospitality & Events

For more information, pricing enquiries or to order a free trial, please contact your local sales team: www.tandfebooks.com/page/sales

 Routledge
Taylor & Francis Group

The home of
Routledge books

www.tandfebooks.com

Printed and bound by CPI Group (UK) Ltd, Croydon, CR0 4YY

21/10/2024

01777055-0007